A Research Primer For Technical Communication

Methods, Exemplars, and Analyses

A Research Primer For Technical Communication

Methods, Exemplars, and Analyses

Michael A. Hughes
George F. Hayhoe

NEW YORK AND LONDON

Microsoft product screen shots reprinted with permission from Microsoft Corporation.

Routledge
Taylor & Francis Group
270 Madison Avenue
New York, NY 10016

Routledge
Taylor & Francis Group
2 Park Square
Milton Park, Abingdon
Oxon OX14 4RN

Routledge is an imprint of Taylor & Francis Group, an Informa business

Transferred to Digital Printing 2010

International Standard Book Number-13: 978-0-8058-6335-2 (Softcover) 978-0-8058-6114-3 (Hardcover)

Library of Congress Cataloging-in-Publication Data

Hughes, Michael A.
A Research Primer for Technical Communication: Methods, Exemplars, and Analyses / Michael A. Hughes and George F. Hayhoe.
p. cm.
Includes bibliographical references and index.
ISBN 978-0-8058-6114-3 (hardcover: alk. paper) -- ISBN 978-0-8058-6335-2 (pbk.)
1. Communication of technical information--Research. I. Hayhoe, George F. II. Title.

T10.5.H84 2007
808'.0666--dc22 2007003449

Visit the Taylor & Francis Web site at
http://www.taylorandfrancis.com

Contents

Preface

Whether you are a student or a practitioner, this book contains essential information you need to know to perform, or to be an informed consumer of, research in the field of technical communication. First, it is a primer for how to conduct and critically read research about technical communication. Second, it reprints four research articles with commentary that analyzes how each article exemplifies a particular type of research report. Last, the content of the articles themselves provides grounding in important research topics in the field.

The book is presented in two sections. The first section discusses the role of research in technical communication and explains in plain language how to conduct and report such research. This section covers both qualitative and quantitative methods, and presents the required statistical concepts and methods in ways that can be understood and applied by nonmathematicians. This section is not intended to be an exhaustive discussion of qualitative and quantitative analysis, but it is sufficient to enable you to conduct research projects and write reports of your findings. The chapters are structured and sequenced to help you identify a research topic, review the appropriate literature, construct a test plan, gather data, analyze the data, and write the research report. Each chapter contains specific activities that will help student readers complete a research project. For practitioner readers, this first section also stresses how to apply its principles as a critical consumer of research articles within the field.

The second section is a collection of articles from *Technical Communication*, the premier journal in the field of technical communication and the official journal of the Society for Technical Communication. The articles have been selected both for their contribution to the body of research in this field as well as for their ability to illuminate the principles of research explained in the first section of the book. Each chapter in this section begins with an introduction that places the subject article in the context of its research methodology and its contribution to scholarship in the field.

People learn best by doing, and this book will help you learn about research by tutoring you through your own research projects and letting you read actual research articles in this field. Activities and checklists at the end of each chapter will help you apply the principles of that chapter to your actual projects. We think that you will find this book an essential how-to if you are a student doing research projects, or as background reading if you are a technical communication professional who reads the literature as a way of staying abreast of your ever-changing field.

Part One

METHODS

1

Research: Why We Do It and Why We Read It

Introduction

If you were asked to envision someone "doing research," you might imagine a student in a library sitting amid stacks of books. Or you might picture a scientist in a white coat sitting in a laboratory surrounded by flasks of colorful, bubbling liquids, peering through a microscope. Certainly, both are examples of people doing research, but they illustrate a narrow, although popular, view about research: namely, that it is academic and somewhat removed from our everyday world.

This book, however, is about research that shapes our professional practice—that is, research that informs practical decisions technical communicators make. The purpose of this chapter is to introduce you to the role that research plays in technical communication and the types of research that can be done. Its goal is to encourage you to think about your own research agenda as a student or practitioner of technical communication, or the ways in which you could apply the research of others to your own practice as a technical communicator. It also tries to help you understand that the way a researcher thinks about and approaches a research project can influence what he or she finds.

Learning Objectives

After you have read this chapter, you should be able to

- Classify research based on its goals
- Classify research based on its methods
- Describe the role of industry, academia, professional communities, and government in regulating the admission of new knowledge into a field of practice
- Describe the different hierarchies of research publications

What is Research?

Peter Senge (1990) uses the term *abstraction wars* to describe a kind of debate typified by a free-for-all of opinions and personal beliefs. You encounter these types of arguments quite frequently among technical communicators (as you would among practitioners of any profession). For example, a group of writers trying to collaborate on a set of documents might argue about what content the users really want or how readers typically use online help. They might also argue about the mechanics of documents, such as what kind of font should be used for the body text, how many steps procedures should be limited to, whether important notes should be printed in boldface, and so forth.

But in a professional field of practice such as technical communication, abstraction wars should not dictate the tenets of the practice. What makes up good technical writing should not rest on arbitrary whims of the individual writer or the personal persuasiveness of those advocating a particular standard or technique. There needs to be a way that the best practices of a profession can emerge as a recognized and reliable consensus among the practitioners of that profession. Well-conducted research can be such a way.

When someone in a meeting says, "Users don't want to go to an online reference to get this information," a reasonable counter is to ask, "What makes you say that?" Typically what the questioner is seeking is evidence or *data*, and beyond that, an indication of how the data supports what the speaker is advocating. What differentiates reasonable arguments from abstraction wars is the use of verifiable observations to support the point being advocated, what the field of action science calls building a *ladder of inference* back to directly observable data (Argyris, Putnam, and Smith 1985).

A Definition

In essence, the linking of actions, decisions, or advocacy to observable data is what research is all about. In this book, the term *research* is used to mean *the systematic collection and analysis of observations for the purpose of creating new knowledge that can inform actions and decisions.* Let us look at what this definition implies:

- Research is systematic—The purpose of this book is to describe a repeatable process for conducting research, one that has protocols and safeguards meant to deliver reliable outcomes.
- Research involves the collection and analysis of data—It is important to note that these are two separate activities. The mindset of the researcher is first to focus on gathering good data and *then* to determine what it means. Data gathered to "prove a point" will almost invariably prove that point, meaning that researchers' preconceptions and biases will influence the research design and the data analysis. Researchers must always be willing to go where the data takes them.
- Research creates new knowledge—Do not be misled by misconceptions from high school, where a "research paper" was intended to show the teacher what you had learned from reading the encyclopedia or from the Internet. In a field of practice, such as technical communication, research should advance our collective knowledge of our field.
- Research should inform actions and decisions—Because technical communication is a field of practice, the outcome of research should enable us to do our jobs better. Research in our field takes on a pragmatic aspect associated with the kind of research often called *action research.* "In action research, a researcher works with a group to define a problem, collect data to determine whether or not this is indeed the problem, and then experiments with potential solutions to the problem" (Watkins and Brooks 1994, 8).

The concept of research also carries with it the assumption that the knowledge created is applicable at a generalized level and is repeatable over multiple instances, producing the same results. For example, it is not the purpose of research to tell us how one particular reader behaves; its value comes from describing how readers in general or a class of readers behaves. Similarly, it would not be the role of research to describe merely how readers used a specific document; its value would come from describing how readers use documentation in general or a genre of documentation. These last examples illustrate one of the primary challenges researchers face: *Research must ultimately articulate generalized truths from specific instances.* For example, if a researcher wants to know how readers in general process information from Web pages, that researcher cannot look at all readers nor analyze all Web pages. The researcher has a narrow access to *some* readers on *some* Web pages and must optimize this limited opportunity to learn as much as he or she can about readers and Web pages in general.

Research in Technical Communication

Research in technical communication is not an activity conducted in a vacuum; it is generally initiated by a problem or a need to understand a phenomenon related to technical communication. Nor is it an activity conducted for its own sake; its conclusions should move the field of technical communication forward, improve technical communicators' decisions, and make their actions based on those decisions more effective than if they had acted without that knowledge. Just as individuals in a meeting want inferences based on data to support someone's assertions, the field of technical communication relies on research to inform best practices within it.

Classifying Types of Research

As much as both researchers and readers would like to believe that research is a totally objective undertaking, it is not. Ultimately, it is shaped by the goals of the researchers and the methods they choose to use. Those who do not acknowledge these influences will not be able to manage their own biases or perspectives as researchers, nor be able to critically evaluate potential bias or the effects of researcher perspective in the research they read. The purpose of this section is to examine the different goals researchers have, the methods they employ, and the ways that those goals and methods can direct and affect the outcome of research.

The Scientific Method

Starting with Francis Bacon in the seventeenth century, research has been associated with a methodology called the *scientific method*, characterized by the following process (Bright 1952):

1. Making physical observations of some phenomenon or aspect of the physical universe
2. Stating a tentative description, called a *hypothesis*, that explains the observations
3. Using the hypothesis to make a prediction about what effect would follow a certain action
4. Conducting an experiment to see whether the predicted effect actually does result from the stated action

This model evolved primarily within the physical sciences and is still widely used. However, with the advent of social sciences such as anthropology and sociology, research has developed a broader set of methods. Although all of these methods rely on the observation of data and on rigorous techniques for validating the conclusions drawn from that data, research is no longer bound to a strict reliance on hypotheses and experiments as defined in the scientific method.

The following discussions describe the more complex landscape of modern research by classifying research genres by their goals and their methods.

Goals of Research

Research goals, in effect, act as lenses that affect how the researcher filters and interprets data, just as tinted lenses not only help a photographer focus on certain details of a landscape but also prevent other details from being emphasized. Therefore, part of the photographer's science and skill is being able to select the correct lens for a specific objective or type of subject. Similarly, part of the science and skill of a researcher or critical reader of research is to select or recognize what goals are driving the research being conducted or studied.

This concept of researcher as filter is an important one and is often overlooked by researchers and readers of research alike. To recount an example, one of the authors of this book was leading a class on a tour of a university's usability lab, and the lab director pointed out the logging station. The director explained that the role of the logger was to keep a running narrative of the user's actions. The author took the opportunity to interject and point out to the students the importance of the logger's role: the logger performed the first filtration of the data; that is, the logger made decisions about what data to log and what not to log. The lab director seemed somewhat abashed by this comment and claimed that their loggers recorded all the data. The author asked whether loggers typically noted details such as the type of shoes the users wore, the color of their hair, details about their complexions, or when they scratched their arms, shifted in their chairs, or brushed back their hair. The director almost snorted and replied, "Of course not." And certainly they should not, but the point is that every researcher applies filters and constantly makes decisions about what data is important and what is not. When researchers do not acknowledge these filters, and when readers of research are not sensitive to the fact that some data has been filtered out before even being shared with the reader, they lose the ability to analyze the results critically.

Categories of Research Goals. To form your own research agenda or to better understand the research done by others, it is helpful to understand the various goals that drive research. Reeves (1998), as the editor of the *Journal of Interactive Learning Research,* identified six categories of educational research goals. Technical communicators face many of the same questions and situations faced by educators, and these classifications can help researchers in technical communication as well. Table 1.1 describes Reeves's classifications in the context of technical communication.

Theoretical Research

Technical communication is a field of practice more than a field of study. As such, not many technical communicators engage in theoretical research, but they often rely on the theoretical research conducted in other fields, such as cognitive psychology or human factors.

Table 1.1 Classifications of research (based on goals)

Research goal	Description
Theoretical	Focuses on explaining phenomena without necessarily providing an immediate application of the findings
Empirical	Focuses on testing hypotheses related to theories of communication and information processing in the context of their application to technical communication
Interpretivist	Focuses on understanding phenomena related to technology's impact on how humans interact with technology, how users interact with the products technical communicators produce, or how technical communicators interact with other roles within an organization
Postmodern	Focuses on examining the assumptions that underlie applications of technology or technical communication, with the ultimate goal of revealing hidden agendas and empowering disenfranchised groups
Developmental	Focuses on the invention and improvement of creative approaches to enhancing technical communication through the use of technology and theory
Evaluative	Focuses on a particular product, tool, or method for the purpose of improving it or estimating its effectiveness and worth

For example, a classic theoretical research article that has had a profound impact on technical communication is George Miller's "The magical number seven, plus or minus two" (1956).* In that article, Miller defines limitations in human capacity to hold information. His observations have influenced many practices in technical writing, such as the optimal way to chunk data so it is easily processed and remembered.

The point is that although Miller's article has heavily influenced technical communication, Miller does not specify concrete practices within it. The purpose of theoretical research is to understand the phenomenon, not necessarily to point to its ultimate application. This seeming lack of an action-based resolution does not belittle the practicality of this kind of research; it merely points out that theoretical research is foundational, often driving the other kinds of research described in the following discussions.

Empirical Research

Empirical research is the type that most people are familiar with, the research typified by the scientific method. It involves the traditional testing of hypotheses that often come out of theoretical research. Empirical research often tests a hypothesis by comparing the results of a control group against the results of a group that has had a particular intervention applied. For example, a technical communicator might read Miller's study "The magical number seven, plus or minus two" and form the hypothesis that the error frequency (when users enter long software identification keys) needed in an installation procedure might be less if the numbers were chunked into groups of five digits. This hypothesis could then be tested by comparing the performance of users who have the long, unchunked software identification key with the performance of users who are presented the chunked key.

The advantage of empirical research is its relative objectivity. Its conclusions are based on quantifiable observations and well-accepted statistical analysis techniques.

* You can find this article at http://psychclassics.yorku.ca/Miller

On the other hand, one of the disadvantages of empirical research is its narrow focus: *what you learn is tightly constrained by the question you ask*. In the previous example, the researcher certainly might learn which way of presenting the software key was better (chunked vs. unchunked), but it is highly unlikely that the researcher would discover an entirely novel approach to securing software that did not involve user-entered numbers.

Interpretivist Research

Interpretivist research is relatively new (compared to theoretical and empirical) and comes from the social sciences, where it was originally termed *naturalistic inquiry* (Lincoln and Guba 1985). The main focus of interpretivist research is to *understand* rather than to test. Hendrick (1983) states that the purpose of this kind of research is to illustrate rather than provide a truth test. Whereas the example about testing the hypothesis of chunked software license keys tested a very specific question ("Is chunked better than nonchunked?"), an interpretive approach would take a more open-ended strategy. Instead of asking, "Is this way better than that way?" the interpretive researcher might ask, "What makes software installation feel easy?" The observations might not be as quantifiable in this case as in an empirical study, consisting more of descriptions of user behavior or transcripts of interviews. The advantage of this more open-ended approach is that it allows for the discovery of unexpected knowledge. The disadvantage, however, is that its nonexperimental approach (that is, its lack of hypothesis testing) can allow researchers and readers alike to apply their own subjectivity and not reach agreement on the validity and reliability of the conclusions.*

Postmodern Research

Postmodern research is typified by a general cynicism about technology and an interest in social or political implications, especially where technology might disenfranchise certain groups. This type of research has gained a greater foothold in the field of education than in technical communication, but there are issues in technical communication that could be attractive to a postmodernist. For example, where empiricists and interpretivists might try to apply information design principles to developing better approaches to electronic voting machines, postmodernists might want to research the impact that high technology in the polling places has on discouraging older or lower-educated voters. Their research might advocate that technology favors the *haves* and the status quo over the *have-nots* or disenfranchised constituents.

Koetting (1996) summarizes the differences among empirical (which he calls *positive science*), interpretivist, and postmodern (which he calls *critical science*) research: "Positive science has an interest in mechanical control; interpretive science has an interest in understanding; and critical science has an interest in emancipation" (1141).

Developmental Research

Developmental research is targeted at producing a new approach or product as its outcome. Accordingly, a lot of developmental research is conducted by companies. In a strong sense, professional associations play an important role in this regard. For example, many articles published in *Technical Communication*, the official journal of the Society

---

* The concepts of "validity" and "reliability" are discussed in later chapters; for now, think of them as describing the "goodness" of the research.

for Technical Communication, are written by practitioners who are sharing discoveries made while working on developmental projects for their companies. Continuing with the example of different kinds of research that could be spawned by Miller's original theoretical research, a technical communicator might conduct usability tests on different approaches to the interface of a help file to discover how different chunking schemes could make user searches easier and more successful. The advantage of developmental research is its emphasis on practical application. The disadvantage is that its conclusions might not be generalized as easily or as broadly as other models.

Evaluative Research

The main difference between developmental and evaluative research is that evaluative research starts with a completed product whereas developmental research is conducted during the design phase of a product. For an evaluation to be research, however, the outcome must be knowledge that can be generalized beyond the immediate product or process under evaluation.

One of the authors of this book had the opportunity to interview Dr. John Carroll* at the beginning of the author's own doctoral studies. Carroll is the founder of Minimalism and the Edward Frymoyer Chaired Professor in Penn State's School of Information Sciences and Technology. In the interview the author asked for Carroll's advice on how he could direct his upcoming doctoral research. Carroll commented that one of the misconceptions people had was the notion that academia did all the research to discover new knowledge, and then business went about applying it. Carroll believed that the more exciting discoveries were being made by businesses, but that they often lacked the time or expertise to fully understand or document why their advances worked. He thought that academic research could add value by helping in this area. The type of research Carroll was suggesting would be a good application for evaluative research. The advantage of evaluative research, as Carroll points out, is that it draws on what is actually being done by the true thought and technology leaders. The disadvantage is that it is more difficult to generalize the lessons learned when a specific product is the focus of the study.

Exercise 1.1: Classifying Research by Goals

This exercise gives you some idea of how varied research can be in the field of technical communication while letting you practice classifying research by goals.

Label the following descriptions of research projects by the type of research goal each seems to be pursuing. Use the following codes:

T Theoretical
EM Empirical
I Interpretivist
P Postmodern
D Developmental
EV Evaluative

* Interview conducted on 10/28/97.

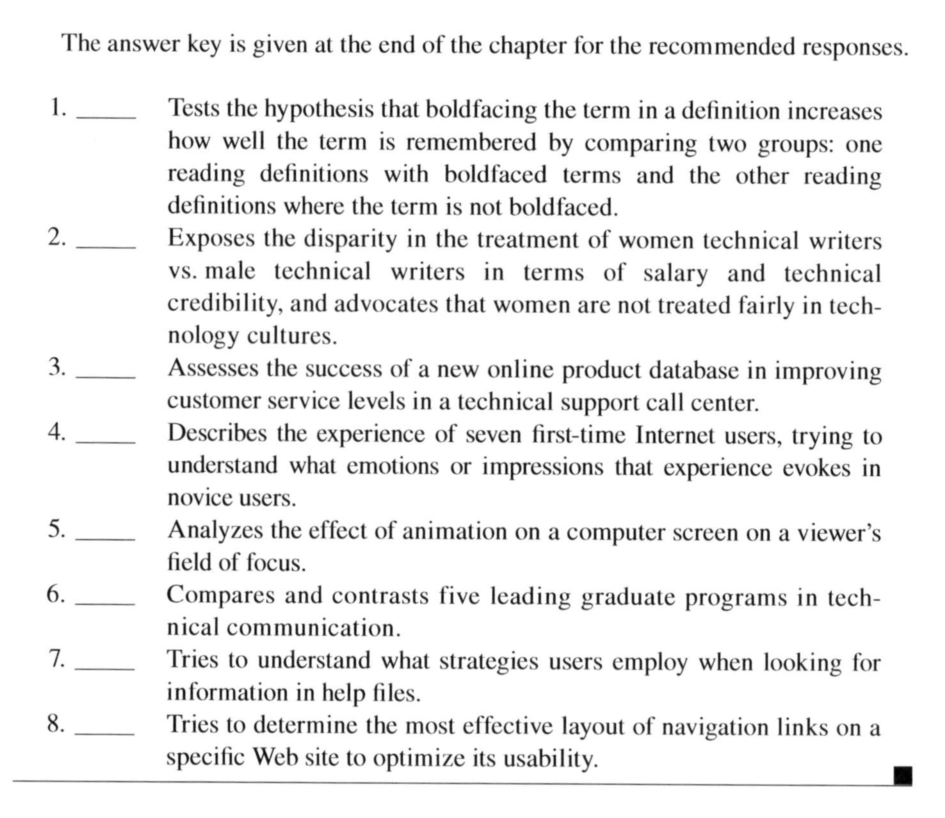

The answer key is given at the end of the chapter for the recommended responses.

1. _____ Tests the hypothesis that boldfacing the term in a definition increases how well the term is remembered by comparing two groups: one reading definitions with boldfaced terms and the other reading definitions where the term is not boldfaced.
2. _____ Exposes the disparity in the treatment of women technical writers vs. male technical writers in terms of salary and technical credibility, and advocates that women are not treated fairly in technology cultures.
3. _____ Assesses the success of a new online product database in improving customer service levels in a technical support call center.
4. _____ Describes the experience of seven first-time Internet users, trying to understand what emotions or impressions that experience evokes in novice users.
5. _____ Analyzes the effect of animation on a computer screen on a viewer's field of focus.
6. _____ Compares and contrasts five leading graduate programs in technical communication.
7. _____ Tries to understand what strategies users employ when looking for information in help files.
8. _____ Tries to determine the most effective layout of navigation links on a specific Web site to optimize its usability.

Methods of Research

In addition to being categorized by its goals, research can also be categorized by its methods. The methodology a researcher chooses to employ can shape the outcome of the research in the same way that tools affect an artisan's output. For example, a woodcarver could be given the same raw material (wood) and objective ("carve a pelican") but given different sets of tools—for example, a chainsaw, a knife, or a set of large chisels—and the outcome would be significantly different, based upon which set of tools the carver used. At a more technical level, the look and feel of a Web site or online help file might differ, even if only in subtle ways, based on the authoring tool chosen. The same applies to research. The outcome of the research can be greatly influenced by the methodology chosen to conduct the research and analyze the data. Researchers and critical readers of research, therefore, need to understand the differences in the methods and how they influence the outcome of the research.

Categories of Research Methods

Table 1.2 lists the five categories described by Reeves (1998) in the guidelines for authors for the *Journal of Interactive Learning Research.*

Quantitative

Quantitative data and its analysis are typically associated with our traditional view of research. Quantitative research relies on statistics to analyze the data and to let the researcher draw reliable inferences from the findings. For example, a research project

Table 1.2 Classifications of research (based on methods)

Method	Description
Quantitative	Primarily involves the collection of data that is expressed in numbers and their analysis using inferential statistics. This is the usual method employed in empirical research involving hypothesis testing and statistical analysis.
Qualitative	Primarily involves the collection of qualitative data (data represented by text, pictures, videotape, and so forth) and its analysis using ethnographic approaches. This method is often used in case studies and usability tests in which the data consists of the words and actions of the test users.
Critical theory	Relies on the deconstruction of "texts" and the technologies that deliver them, looking for social or political agendas or evidence of class, race, or gender domination. This method is usually employed in postmodern research.
Literature review	Primarily involves the review and reporting on the research of others, often including the analysis and integration of that research through frequency counts and meta-analyses. This method can be applied to any research goal that aims at integrating prior research.
Mixed methods	Research approaches that combine a mixture of methods, usually quantitative and qualitative. Mixed methods are often found in research involving usability tests, which are a rich source of quantitative data, such as time to complete tasks or frequency of errors, and qualitative data, such as user comments, facial expressions, or actions.

that compared the average times to complete a task using two styles of online help would use quantitative methods (the capturing of data in numerical format and the analysis of that data using statistics).

Qualitative

Qualitative methods rely on nonnumerical data, such as interviews with users or video recordings of users trying to do tasks. Qualitative data is harder to analyze in some respects than quantitative data because it can be more susceptible to subjective interpretation. An example of a qualitative approach would be a field study in which the researcher observes workers using a new software product and takes notes on how they use the documentation to learn the new software. The researcher might also supplement the observations with interviews. The resultant data, then, would consist of notes and transcripts that would be analyzed to see whether meaningful patterns emerged.

Critical Theory

Critical theory looks closely at texts—that is, formal or informal discourses—to determine what they are "really saying" (or what they deliberately are *not* saying) and compares them with their superficial meanings. For example, a study might examine how technical communicators handle known product deficiencies in user guides by analyzing their language styles to see whether stylistic devices such as passive voice or abstraction obscure the meaning of the text and avoid frank discussion of the deficient features.

Literature Review

Although all research projects contain a literature review, some projects are exclusively the review of other research projects that have been published (usually referred to as *the literature).* The purpose might be to look for trends across the research or to collect in one document a cross section of the important opinions and findings regarding a particular topic. For example, a researcher might do a literature review that looks at the relevant research that has been published about the usefulness of readability measures and try to draw conclusions about the trends or findings in general. This kind of review can be particularly helpful to practitioners who want to adopt best practices.

Mixed Methods

It is not unusual to find more than one of the preceding methods being employed in the same research project. For example, a usability test might use quantitative methods to determine where in a Web site the majority of users abandon a task and then employ qualitative methods to understand why. Another example would be an evaluation study that uses interviews to determine which section of a manual is considered the most important by users and then compares how rapidly information from that section can be retrieved compared to a similar section in a competitor's manual.

Exercise 1.2: Classifying research by method

This exercise has you apply the categories just discussed in the context of actual technical communication topics.

Label the following descriptions of research projects by the type of method each seems to be applying. Use the following codes:

QN Quantitative
QL Qualitative
C Critical Theory
L Literature Review
M Mixed Methods

The answer key is given at the end of the chapter for the recommended responses.

1. _____ Interviews graduates of technical communication degree programs after their first year out of the program to see how they feel their education has affected their professional growth.
2. _____ Records the time it takes each of 12 different users to install a product and calculates the average installation time.
3. _____ Examines the wording of the online ads for technical communicators to uncover how age-discrimination messages are being communicated.
4. _____ Counts the number of times users go to the help file and interviews each user to understand why they went to the help file at the particular times they did.

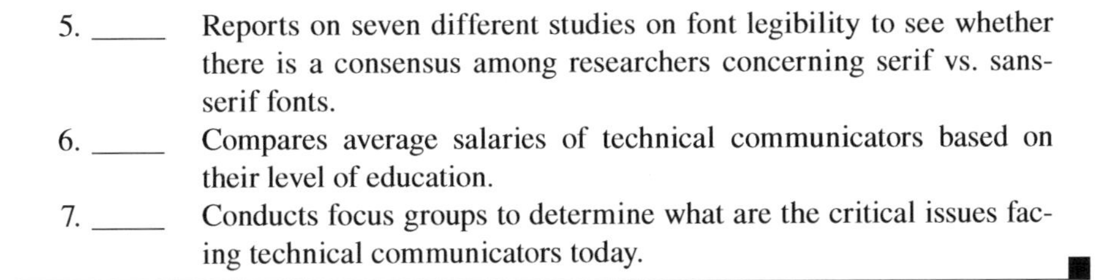

5. _____ Reports on seven different studies on font legibility to see whether there is a consensus among researchers concerning serif vs. sans-serif fonts.
6. _____ Compares average salaries of technical communicators based on their level of education.
7. _____ Conducts focus groups to determine what are the critical issues facing technical communicators today. ■

Table 1.3 Common associations of goals and methods

Goals	Methods
Theoretical	Quantitative
Empirical	Quantitative
Interpretivist	Qualitative
Postmodern	Critical theory
Developmental	Quantitative, qualitative, or mixed methods
Evaluative	Quantitative, qualitative, or mixed methods

Association of Goals and Methods

There is a strong association between the goal of the research and the methods a researcher employs. Table 1.3 shows associations often encountered in research.

If you are a student researcher or an experienced practitioner starting your first formal research project, the choices can seem overwhelming. But your personal interests or work requirements will dictate the goal, and most technical communication research will rely on quantitative or qualitative methods, both of which are covered in this book.

As a critical reader of research, one question you should ask is whether the researcher has applied the appropriate method for the stated or implied goal of the research. A mistake that researchers sometimes make is to choose the method of their research before they have clarified their goal. Often this is due to a preference for qualitative over quantitative (by those afraid of the "number crunching") or vice versa (by those who feel that only measurable phenomena can be trusted). The better approach is to classify the goal first and then to choose the appropriate method. Classifying the goal is a direct outcome of choosing the research topic and research question, which is discussed in the next chapter.

Student Research Activity: Brainstorming Research Topics

The purpose of this activity is to get you thinking about all the possibilities for a research project that you might conduct.

Step 1: For each type of research goal presented in this chapter, identify one to three examples of possible research topics within the field of technical communication. Draw on topics from your studies or real-life problems you deal with as a technical communicator or user of technical documents.

Step 2: For each topic identified in step 1, discuss which method or methods could be appropriate, and give a general example of what kinds of data might be gathered.

Research Sources

Earlier in this chapter, we stated that research in technical communication helps inform best practices within the field. But how does that happen? How does that research get sponsored and disseminated? In effect, there are four sources of research:

- Industry
- Academia
- Professional societies
- Government

The Role of Industry

One of the great contributions that industry makes is that it keeps research in a field relevant and practical. New ideas are scrutinized against two criteria: (1) Is it worth the cost? (Would users pay for it or would it reduce the price of products or services?) and (2) Can you really do it? (that is, get the results you promise). Both of these criteria drive the need for the collection and analysis of data—research. The disadvantage is that industry lacks both the motivation and general mechanisms for distributing the new knowledge it creates.

The Role of Academia

Academic programs in technical communication are another source of research in this field. They have the advantage of having professors who provide both continuity and expertise combined with students who can provide the mental (and sometimes physical) labor. If you are a student getting ready to embark on a required research project as part of your studies, you are part of this mechanism that keeps the field of technical communication vibrant and relevant. Faculty and students also have easy access to conference proceedings and journals, which provide a mechanism for publishing results, which find their way into courses on technical communication and eventually into practice as students enter the workforce.

The Role of Professional Societies

Another source of support for research is the professional societies associated with technical communication. These organizations often sponsor research directly, but most importantly, they provide journals that publish and disseminate the results of relevant research to their members. One such journal is *Technical Communication*, the quarterly journal of the Society for Technical Communication. It is from this journal that the sample research articles for this book have been selected.

The Role of Government

Local, state, and federal governments are still another source of support for research in technical communication. For example, governments might want to gauge the effectiveness of a communication campaign promoting safe sex or evaluate the usability of a Web site that disseminates information about health issues. Government funding of technical communication research is often embedded in grants to natural or social scientists, information technologists, or engineers.

Hierarchies of Publications

Anyone who wishes to have his or her ideas or research published or who wishes to be a critical reader of ideas and research in a professional field needs to understand the different hierarchies of publications and their requirements for publication. There are three levels:

- Open publications
- Editor-regulated publications
- Refereed journals

An *open publication* is one in which anyone can publish without the scrutiny of anyone else as to the validity or reliability of their assertions. The Internet is the largest example of such a medium. Anyone can create and post a Web site to the Internet without anyone's permission or review. Another example of an open publication is called the *vanity press*. In these venues, the author or a company pays all the publication expenses. Another category of open publication is the *white paper*, a document written and published by companies as marketing tools to promote their technologies or processes, or to share the results of their research in the hope of attracting future business. Open publications should be read with a high degree of skepticism because no filtering for adherence to standards of research has been applied by third parties.

An *editor-regulated publication* is one in which the editor or an editorial staff decides which submissions will be published. Criteria such as interest of the subject matter to readership, reputation of the author, and quality of the writing are often applied. Examples of editor-regulated publications are newsletters and magazines. These publications often have an informal style and do not rely on footnotes or references to other research to support the author's assertions.

Refereed journals represent the most rigorous screening of submissions for publication. Submissions are initially evaluated by the editor and then referred to an independent review committee recruited for the purpose of evaluating the manuscript. These reviews are called "peer reviews" because they are conducted by practitioners or academics who share the author's level of expertise (or an even higher level of expertise). These reviewers critique the article, often requiring that the author elaborate on or reevaluate assertions made in the article. Sometimes, enthusiastic researchers make unwarranted assertions or claims about their results, and peer reviewers often help them recast their claims into a more conservative and more reliable scope. Often, these peer reviewers will reject the submission for not meeting the standards of reliable research. For these reasons, research articles that come from refereed journals carry higher credibility and reliability than ones appearing in open or editor-regulated publications.

Some book publishers—university presses, for example—combine the methods of editor-regulated publications and peer-reviewed journals. In these cases, an editor makes an initial judgment about the quality of the work, and then all or portions of the manuscript are subject to peer review.

Summary

Research is the systematic collection and analysis of observations for the purpose of creating new knowledge that can inform action. Research in technical communication informs practical decisions technical communicators make. Its value is that it shapes the best practices of the field.

Research can be categorized by its goals:

- Theoretical
- Empirical
- Interpretivist
- Postmodern
- Developmental
- Evaluative

Research can also be categorized by its methods:

- Quantitative
- Qualitative
- Critical theory
- Literature review
- Mixed methods

The publications that disseminate research can be sorted into three categories:

- Open publications—No filtering or selection process.
- Editor-regulated publications—Submissions are selected by the publication's staff.
- Refereed journals—Submissions are peer-reviewed not only for the quality of the writing, but for the rigor of the research methods employed as well.

References

Argyris, C., R. Putnam, and M.M. Smith. 1985. *Action Science: Concepts, Methods, and Skills for Research and Intervention*. San Francisco, CA: Jossey-Bass.

Bright, W.E. 1952. *An Introduction to Scientific Research*. New York: McGraw-Hill.

Hendrick, C. 1983. A middle-way metatheory. *Contemporary Psychology* 287:504–507.

Koetting, J.R. 1996. Philosophy, research, and education. In *Handbook of Research for Educational Communications and Technology*, ed. D.H. Jonassen, 1137–1147. New York: Simon & Schuster Macmillan.

Lincoln, Y.S., and E.G. Guba. 1985. *Naturalistic Inquiry*. Beverly Hills, CA: Sage.

Miller, G. 1956. The magical number seven, plus or minus two: Some limits on our capacity for processing information. *The Psychological Review* 63:81–87.

Reeves, T. 1998. The scope and standards of the *Journal of Interactive Learning Research*. http://www.aace.org/pubs/jilr/scope.html: Association for the Advancement of Computing in Education.

Senge, P. 1990. *The Fifth Discipline: The Art and Practice of the Learning Organization*. New York: Doubleday.

Watkins, K.E., and A. Brooks. 1994. A framework for using action technologies. In *The Emerging Power of Action Inquiry Technologies*, ed. A. Brooks and K.E. Watkins, 99–111. San Francisco, CA: Jossey-Bass.

Answer Key

Exercise 1.1

1. Empirical. The key words *tests* and *hypothesis* point right away to a scientific method approach, placing this study squarely in the empirical category.
2. Postmodern. The emphasis on social disparity, in this case focusing on gender inequity, coupled with a definite point of advocacy, that is, women are not being treated fairly, makes this a postmodernist study. In effect, it is pointing out how technology is being applied to a group's disadvantage—a common theme in postmodernist research.
3. Evaluation. The results of a specific product are being evaluated. Although the study would look at one instance, it could lead to the generalization that online databases might improve customer service levels in other instances.
4. Interpretivist. This study is trying to understand a general phenomenon, looking to learn how people experience or make sense of a technology experience. It has no prior assumptions, hypotheses, or political agenda. Interpretivist studies can be recognized by this open-ended approach of "What happens when" or "How do people makes sense of" to technical-communication-related experiences.
5. Theoretical. The clue here is that the root of the question is in human psychology or physiology. The findings are likely to be broad descriptions of how human sensing or mental processing works. Its findings would not lead to a specific outcome as much as they would lead to additional research that applied its findings to specific designs or processes.
6. Evaluative. In this case, specific programs are being compared and contrasted. The outcome of this study could be applied to making a decision about which program to attend or its insights could be useful to someone contemplating starting such a program within his or her own university.
7. Interpretivist. The emphasis is on trying to understand how users make sense of a technical communication task, in this case, using help files. It is open-ended and does not purport to test a theory.
8. Developmental. The researcher is trying to optimize a specific product or design. Even so, its findings may point to a best practice that can be applied to other designs; thus, it is included in the field of research.

Exercise 1.2

1. Qualitative. The data being taken and analyzed consists of interviewees' words.
2. Quantitative. The data consists of numerical measurements.
3. Critical theory. Even though the data consists of words, the analysis is looking for hidden meanings behind the text.
4. Mixed methods. The study uses quantitative (number of times users go to help) and qualitative (interviews to understand why) data.
5. Literature review. The study is a *study of studies* and not an original research project.
6. Quantitative. The data is all numeric, i.e., salaries and levels of education.
7. Qualitative. The data is the words and other forms of expression gathered or observed in the focus group.

2

The Research Phases and Getting Started

Introduction

This chapter introduces the components of a formal research report so that as a researcher you understand what you are trying to produce, and as a reader of research, you have some landmarks to help navigate through your reading. It also provides a sample project plan for helping you plan and track a research project.

This chapter also explains an important requirement for conducting research: gaining the informed consent of the participants.

Finally, this chapter helps the student or first-time researcher tackle the beginning of a project, namely, defining the research goal and articulating the guiding research questions.

Learning Objectives

After you have read this chapter, you should be able to do the following:

- Describe the major sections of a formal research report
- Plan a research project
- Describe how to protect the rights of human subjects
- Write a research goal statement
- Write appropriate research questions

The Structure of a Formal Research Report

Before undertaking any project, it is useful to have a picture of what the finished outcome should look like. Not only is this true for assembling bicycles and barbecue grills, but it applies to research projects as well. The outcome of a research project is a research report, a formal document that describes how the research was conducted, summarizes the results, and draws meaningful conclusions. It is often the basis of a research article that gets published in a journal.

Not every research paper will contain all the elements of a formal research report, and articles are often trimmed down. Even so, the following structure should serve as a foundation for reporting research, and its elements should be reflected to some degree in every research report and article.

A formal research report contains the following sections (Charles 1998):

- Statement of the research problem
- Review of the literature
- Description of the methodology
- Analysis of the data
- Conclusions

In a large report, such as a doctoral dissertation or master's thesis, each of these sections is a separate chapter. In shorter reports, they can be sections with their own headings. Often, in articles, the sections can be as small as just a few paragraphs and might not be marked as sections with headings. Still, a good research report reflects these elements, and a critical reader of research articles should expect them and look for them in whatever form they take.

Statement of the Research Problem

The section or chapter of a research report that states the research problem is often called the *introduction*. Its purpose is to provide the context for the research, in essence, why the researcher thought it was an important topic and what in particular he or she set out to learn.* In writing the introduction, the researcher needs to focus on three areas:

- Background—In general, the researcher needs to describe what problems or activities in the field of technical communication the reader needs to be aware of to put the research topic into context. In short, why should the reader care about this research?
- The goal of this particular research—The researcher focuses on a specific aspect of the general problem area that this research project tries to address.
- The research questions—The researcher identifies the specific questions, i.e., what he or she will observe or measure to meet the research goal.

For example, a researcher might want to conduct research related to font selection. The background section could discuss that there are many fonts now available to technical communicators and explain that font selection is often hotly debated during the design phase of a document. This establishes that the topic is current and has importance for technical communicators.

Next, the researcher narrows the scope of the research by stating a specific goal for the research project. In the font example, it might be: Determine whether font selection affects reader effectiveness when using online help. Now the reader has a better idea of what aspect of the general topic the researcher intends to investigate. But the kind of information or type of data the researcher will collect in trying to meet that goal is still unclear.

The third component, the research questions, helps clarify how the researcher hopes to achieve his or her goal; that is, what specific aspects or phenomena the researcher

* It was no accident that we said "set out to learn" and not "set out to prove." A good researcher brings an open mind to every project and must be willing to be surprised by his or her findings if that is where the data leads. Also, the results of research seldom definitively "prove" a conclusion.

will observe or measure. In the font example, the researcher might specify the following research questions:

1. Does choice between serif and sans-serif fonts in instructions affect time to complete task?
2. Does choice between serif and sans-serif fonts in instructions affect the frequency of errors when following online instructions?
3. Does choice between serif and sans-serif fonts affect user perception of the difficulty of the task?

Note that the research questions are what actually shape the character and scope of the research. In the previous example, the research study will probably go on to be empirical and quantitative because it asks quantitative questions such as how fast and how many. However, it could end up being a mixed-methods study, depending on how the researcher decides to deal with the third question. The researcher may decide to deal with perceived difficulty *quantitatively* by measuring user perception on a scale or *qualitatively* by using observations and interviews as the source data.

In a quantitative study, the researcher often states the test hypotheses as well. These are derived from the research questions and state the assumptions that will be tested, usually naming the intervention, the control group, and the test group. For example, one of the hypotheses from the earlier sample questions on fonts might be: There is a statistically significant difference in the average time to complete a task made by readers who use help files displayed in sans-serif fonts compared to users who use the same help content displayed in serif fonts.

We discuss how to write test hypotheses in the chapter on quantitative methods.

Review of the Literature

Research reports are filled with references to other articles or books that the researcher has consulted as part of the study. In the introduction, for example, the researcher will probably refer to several sources in making his or her case that the topic is of importance or in giving the reader the necessary background information to understand the context of the study. Even so, there is usually a separate section of the report devoted specifically to reviewing certain kinds of literature, and in a formal report this section is called the *literature review*. We have an entire chapter in this book dedicated to the literature review, but we include a brief discussion here to put this important section in context of the overall structure of the research report.

The purpose of the literature review is threefold:

1. Establish the scope of prior research—The purpose of new research is to *advance* a field's understanding of a topic. To accomplish this goal, the researcher must be familiar with what has been done already. A research advisor would not let a student just decide to "see whether font selection has an effect on readability" and set off to gather data. The advisor would point out that others have probably already looked into this question and insist that the student start by studying what has been learned so far. The literature review

tells the reader what the researcher found in reviewing what others had already done in the area being studied.

2. Educate the reader—The literature review often presents a comprehensive review of the topic being researched so that the reader can more fully understand the context of the findings the researcher will report. The introduction started this reader education process by establishing the importance of the topic within the field of technical communication, and the literature review continues the process in more depth. Often, the literature review presents the technical background the reader would need to understand the research. For example, in a study about font selection, the literature review could contain references from articles that define the key attributes of font design, such as serif vs. sans-serif, x-height, weight, size, and so on.
3. Ground the researcher's premises in data—People carry a lot of assumptions with them into a discussion of any topic. Some of those assumptions are valid, and some are not. Researchers are no different. Part of the purpose of the literature review is to challenge these assumptions by expecting the researcher to support them with data. For example, a statement such as "Users want an interface that is easy to use" in a product design meeting may not elicit even token resistance from any of the practitioners in the room. The same sentence in a research study could evoke the response, "Says who?" from a critical reader of the article. It is not that we do not believe it; it is just good protocol to ask that the researcher support all the assertions and assumptions that he or she brings into a study. That particular assertion, for example, could have been supported by citing studies that showed products rated as user-friendly are more successful commercially than those that are not, or by citing studies that showed online documentation rated as easy to use was referred to more often than documentation rated difficult to use. If, in fact, a basic assumption you want to make cannot be supported by prior research, testing that assumption might be the more appropriate topic for your study. This discipline is similar to going through security at the airport: We x-ray all the carry-on bags to screen out the dangerous ones. In research, we question all the assumptions to screen out the invalid ones. In this regard, the literature review is an important discipline to help researchers rid themselves of biases and unfounded assumptions they might bring into the study.

Description of the Methodology

Once the researcher has introduced the background and reported on the literature in the area of interest, the next task is to tell the reader how the study was conducted.

Name the Method Type. The researcher should explain what research method was chosen and why:

- Quantitative
- Qualitative
- Critical theory
- Literature review
- Mixed-methods

As a critical reader of research, you should ask yourself whether the researcher's method matches the goal and research questions. For example, if the research question was, "Why do users go to online help?" and the method chosen was quantitative—for example, count how often that users open the help file—the critical reader should question the researcher's choice of methodology. A qualitative method that looked to illuminate user motives would have been a more appropriate choice.

As a researcher yourself, you should carefully select the method based on the goal and the question—not on your individual comfort level with a particular method or the ease of applying it.

Describe the Data Collection Instruments or Events

Closely linked to the method type is the actual way the researcher gathers data. The following list is just a sample:

- Case studies
- Surveys
- Interviews
- Focus groups
- Usability tests
- Discourse analyses (analyzing text)
- Controlled experiments (for example, bringing people into a formal lab and measuring performance such as speed to complete a task)
- Analyses of existing data (for example, analyzing salary data reported by the Society for Technical Communication to see whether there are trends or relationships)

Describe the Sample and the Sample Selection Process

Unless the study is collecting data on the entire population of interest (which is unlikely), the data will be derived from and therefore represent only a sample of that population. The methodology section should explain how the researcher went about selecting that sample and why it is a valid representation of the population as a whole. As a researcher, you should articulate any concerns you might have in that regard and how you have tried to manage possible shortcomings. Doing so will add to the credibility of the study.

For example, a student might choose to conduct a survey of fellow students in his master's program because they are easy to get to and perhaps more willing to participate than others. Such a sample of the entire population is called a "sample of convenience." It is acceptable to choose such a sample, but the good researcher will note that fact and warn the reader that the attitudes or behaviors of postgraduate students in a field might not be indicative of the full range of practitioners in the field. No researcher ever has access to a perfect data collection environment; that fact does not stop us from conducting research. Being honest with yourself and your reader about the limitations or constraints you have had to manage, however, will add value and credibility to your findings.

Describe the Data Analysis Techniques

Last, the methodology section should provide an overview of what method was used to analyze the data. We talk about data analysis techniques in detail later in this book. But for now, know that in the methodology section the researcher should describe at a high level how the data was analyzed.

If you are doing an experimental and quantitative study, it is customary to state your test hypotheses and the statistical tests you applied. If you are doing a qualitative study, you need to describe the general methods you used to analyze the qualitative data.

Analysis of the Data

The theme of the data analysis section in a research report is, "It is what it is." As a researcher you need to distance yourself subjectively as much as you can in this section from judgmental statements, conclusions, or recommendations. For example, the data might say that Help File A resulted in users performing a task quicker, with fewer mistakes, and with higher satisfaction ratings than with Help File B. Does this mean that Help File A is better or preferred? Although you may well have come to that conclusion, that is something you should not talk about in the data analysis section. It might seem unnecessary to do this in a practical environment, but remember that research is designed to separate truth from biases and assumptions. Research is a discipline that tries to ensure that our conclusions and recommendations are driven by the data and not by whatever preconceptions we came into the study with. For this reason, it is good protocol to first review what the data state before drawing conclusions or calling for specific recommendations. The fact that the readers will start to draw their own conclusions about the superiority of Help File A will add credibility to your conclusions section when you draw that same conclusion.

Conclusions

In the final section of the report, the voice of the researcher is allowed to emerge. You have done your due diligence in the literature review by grounding your study in what those who have gone before you have done and said, and you have kept your judgments and opinions out of the data analysis section. Now, in the conclusions section, you may speak your own mind and voice what the data means to you.

A word of caution, however: Do not get too far out in front of your data. For example, if you tested to see how readers used indexes, and you based your study on a sample of convenience using students from a master's program in technical communication, avoid drawing broad conclusions about how "users" use indexes or the unwarranted claim that the index approach tested is "superior to conventional indexing methods." People in general might not read the way technical communication graduate students read.

You would also have to temper your findings based on the kinds of documents that were used in the study. For example, a study that used software reference manuals might not tell us how people would use a book on automobile maintenance. One of the more common researcher errors is to overstate the application of a study's findings. Our advice: Do not overreach; be satisfied to add a little to the general knowledge of a subject.

Exercise 2.1

Read "How the Design of Headlines in Presentation Slides Affects Audience Retention" in Chapter 8 and identify the sections of a research report we have discussed in this chapter. ■

The Phases of a Research Project

The actual phases of conducting a research project very closely parallel the structure of the final report:

- Identifying a research goal
- Formulating the research questions
- Reviewing the literature
- Designing the study
- Acquiring approvals for the research
- Collecting the data
- Analyzing the data
- Reporting the results

A common mistake is to think that you write the research report at the end of the study. Actually, each section of the report is meant to help guide the study, and the report should start to emerge from the very outset of the project. It will be an iterative process, for sure. You will start the introduction, read some literature and start the literature review, then gain a better understanding of the problem, and revise the introduction, etc. So, at some times the report points you forward, and at other times it documents where you have been. It is a living document during the study; feed and nurture it often.

In a very formal study, such as a master's thesis or doctoral dissertation, or in the proposal for a research project for which you are seeking funding, the researcher must often write a proposal or research prospectus before proceeding with the study. In this case, the first three sections—that is, the statement of the research problem, the review of the literature, and the description of the methodology—serve as the body of the proposal or prospectus.

Even if a formal proposal or prospectus is not required, the researcher should draft these three sections before actually gathering any data. Often, the insight that comes from the literature review or the planning of the data analysis help drive what data needs to be gathered and how the data is gathered. A common mistake among new researchers is to gather data (for example, send out a survey) before they know what their goal and research questions are or how they intend to analyze the results. In short, they go looking for answers before they understand the questions.

In general, if you are writing a research proposal or prospectus, the methodology section is written in the future tense—telling the reader what you intend to do. If you are writing a research report or article, the methodology section is written in the past tense—telling the reader what you did.

Managing the Research Project

Figure 2.1 shows a 12-week project plan for conducting a research project. Some research projects can take less time, some more. A doctoral dissertation, for example, can require several years to complete. The rectangles represent tasks, and the diamonds represent milestones (deliverables against which you can measure your progress). In this case, the project plan uses the sections of the report as the milestones. This approach helps ensure that you keep your documentation up to date as you proceed through the project.

Bear in mind that real projects tend to circle back onto themselves (Smudde 1991); that is to say, although you will have "completed" a draft of your introduction

Figure 2.1 Twelve-week plan.

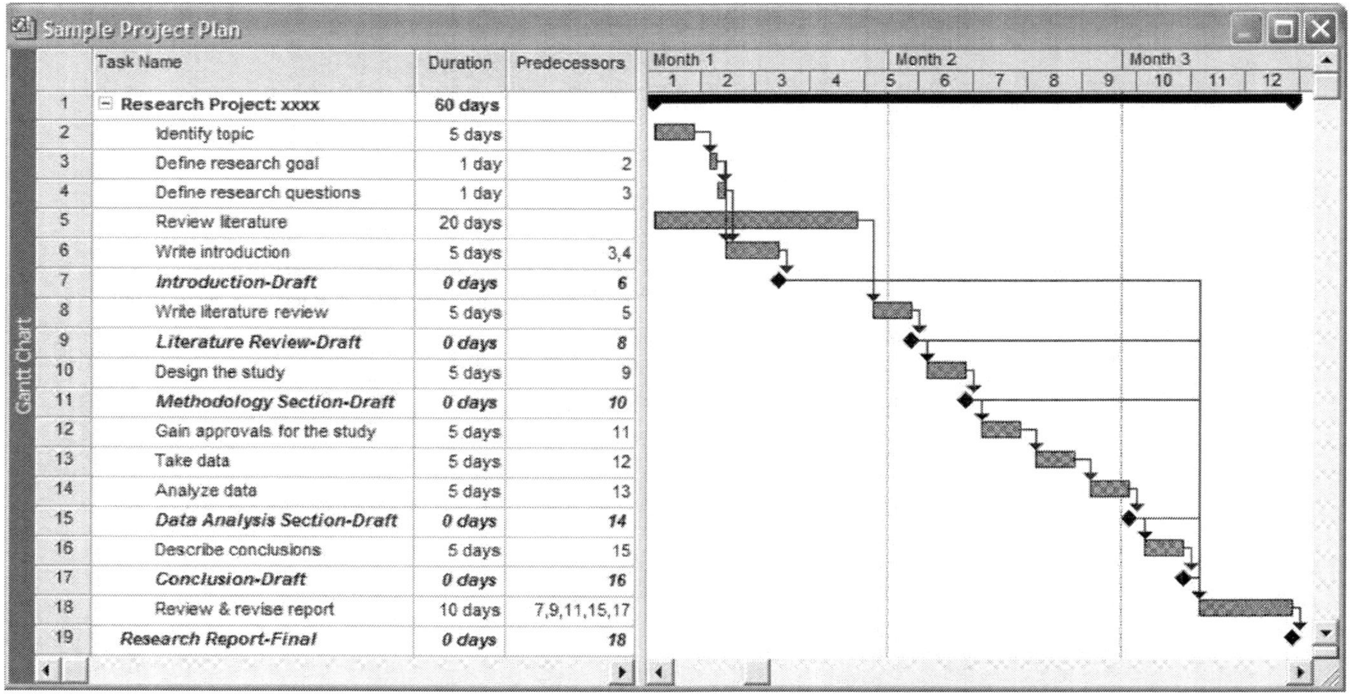

Sample Project Plan

	Task Name	Duration	Predecessors
1	**Research Project: xxxx**	**60 days**	
2	Identify topic	5 days	
3	Define research goal	1 day	2
4	Define research questions	1 day	3
5	Review literature	20 days	
6	Write introduction	5 days	3,4
7	***Introduction-Draft***	***0 days***	***6***
8	Write literature review	5 days	5
9	***Literature Review-Draft***	***0 days***	***8***
10	Design the study	5 days	9
11	***Methodology Section-Draft***	***0 days***	***10***
12	Gain approvals for the study	5 days	11
13	Take data	5 days	12
14	Analyze data	5 days	13
15	***Data Analysis Section-Draft***	***0 days***	***14***
16	Describe conclusions	5 days	15
17	***Conclusion-Draft***	***0 days***	***16***
18	Review & revise report	10 days	7,9,11,15,17
19	***Research Report-Final***	***0 days***	***18***

before you start to collect data, you may gain a better understanding of your research question during the data-gathering phase and want to make some significant revisions or additions to the introduction. This could, in turn, lead you to reopen your literature review. Technical communicators should be used to this circling back; we often do not fully understand the product we are documenting or the users' needs at the beginning of a project, and we must often revise early chapters of a manual to accommodate our expanded understanding gained from researching and writing later chapters. Be prepared for the same thing to happen in your research projects.

The project plan shown in Figure 2.1 tries to accommodate this expanding understanding the researcher acquires by scheduling the literature review to coincide with and then extend beyond the introduction section. This sequencing recognizes that there are actually two phases to the literature review: (1) a broadly scoped phase in which the researcher is still trying to identify or understand a general topic and (2) a narrower phase in which the researcher is focusing in on literature directly related to the research questions. There is also a final task for revising the entire report. This last task lets the researcher apply his or her final understanding of the research topic across the entire report.

Gaining Permissions

It is very important that a researcher respect the rights of those who are affected by the research. Ensuring those rights often means getting permission or approval to conduct the research. There are three possible tiers that a researcher should consider and possibly seek permission from:

- Organizational sponsors
- Individual participants
- Institutional review boards (IRBs)

Sponsors

If you are going to do research within an organization, for example, a specific company, it must be with the knowledge and approval of that company. Often the manager of the department or division within which the research will take place will suffice, but you should be certain to determine in advance what level of authorization your company requires. It is important that researcher and sponsor agree upon the conditions—for example, will employee time be taken away from productive tasks (such as asking them to participate in interviews or surveys) and will the sponsor be identified in the report.

Sponsor permission is important for consultants who wish to publish the results of a particularly interesting assignment as a case study. If you are a consultant, it is wise to include such a provision in your contract, make sure that the client is comfortable with that intention, and protect any proprietary knowledge assets that rightfully belong to the client. The same caveats apply to employees doing research within the companies that employ them.

Individual Participants

Many research projects involve working with people who are the source of the data, and it is important that the rights of those people are respected during a research project. Biomedical researchers have developed stringent guidelines and rules of ethics to protect human subjects who are participating in research that can directly impact their health and welfare. For them, protecting human rights includes protecting the subject from harm, ensuring that the benefits of the research justify its risks, advocating that those groups or populations chosen to participate in the risks of the research will be among those who realize its benefits, and ensuring that those who participate in research do so willingly and knowingly (Office of Human Subjects Research 1979).

Technical communication researchers should hold themselves to the same standards. Unlike biomedical research, research in technical communication will seldom expose participants to risk. However, the concept of *informed consent* is still an important ethical requirement in research in technical communication.

Informed Consent

The principles of informed consent are ingrained in our modern concepts of ethical research. The Nuremberg Code (1949), the first internationally accepted regulation of ethical research, made it the foremost principle: "1. The voluntary consent of the human subject is absolutely essential" (181). The Belmont Report (Office of Human Subjects Research 1979), the official policy paper of the Office of Human Subjects Research, says that respect for persons requires that subjects be given the opportunity to choose what shall or shall not happen to them. The Belmont Report stipulates three components for informed consent:

- Information
- Comprehension
- Voluntariness

Information

Let participants know what the research is about, what they will be asked to do (or what will be done to them), what risks they might incur, and what will happen with their data. For example, in a research project involving usability testing, the participants should be told that they are part of a research project to understand how people use a software product, that they will be asked to do some typical tasks with that product, that they will be observed, and that their words and interactions with the product will be recorded. They should also be told how any video or audio will be used if their interactions with the product will be taped. For example, if the tape might be used in a highlight video shown at conferences or training workshops, the participants should be told that fact. Above all, participants should be told whether their identity will remain anonymous (not known to the researcher as in an online survey), confidential (known to the researcher but not revealed), or public.

Confidentiality is generally maintained by not using the participants' names or identifying characteristics in the published report. You need to be particularly careful in the area of a participant's *identifying characteristics*, that is, descriptions that will reveal the identity of the participant. For example, a survey that asks for company name and job title will not be anonymous if the participant is the only person in the company with that title, nor will it be confidential if those characteristics are linked to the reported findings. For example, if Jane Smith is the only technical communication manager at ABC, Inc., specific results or comments attributed to "a technical communication manager at ABC, Inc.," are hardly confidential.

Comprehension

It is not enough that you inform the participants. You must also ensure that participants understand what you are telling them. This could be an issue when you undertake research involving minors or where there could be language barriers between the researcher and the participant.

Voluntariness

Voluntariness is essentially the "consent" component in "informed consent." The principle of voluntariness goes back again to the Nuremberg Code: "During the course of the experiment the human subject should be at liberty to bring the experiment to an end if he has reached the physical or mental state where continuation of the experiment seems to him to be impossible" (1949, 182). In short, the participant must be empowered to say no at any time during the research, including not participating at all. Where the researcher has authority or power over potential participants—for example, as a teacher over students or a manager over employees—great care must be taken to ensure that coercion does not come into play.

A teacher, for example, who wants to use certain class projects in a research project, might ask a student to hold onto informed consent forms until the course is over and grades have been issued; that way, the teacher does not know who has agreed to have their work included in the research until after the course has been completed, and students do not have to worry that their decisions might influence their grades.

When Is Informed Consent Required?

What if a research project included tracking the percentage of visitors to an organization's Web site who registered for that organization's newsletter? Would the researcher have to get the informed consent from each visitor? No. Although the decision of when informed consent is required can get complicated, here are three guiding criteria to consider:

- Are the participants at risk?
- Will the participants' words or identities be used as data?
- Is it research or just routine evaluation of what would have occurred normally?

Are the Participants at Risk?

If your research causes risk to the participant above what would ordinarily be incurred in their routine environment, you must get the participant's informed consent. Normally, this is not an issue with the kinds of research technical communicators do. True, users can get frustrated during usability tests, but users routinely get frustrated by software. But if a test is likely to cause unusual stress or anxiety, participants would have to be informed of that probability and would have to give consent.

Will the Participants' Words or Identities Be Used as Data?

People own their words. If you wish to use people's words or equivalent personal expression—such as videotape of their reactions, descriptions of their interactions, and so forth—or if you identify the participants' identities, you need to get their informed consent. The only exception is if the words or expressions already exist in a public domain of some type. For example, you do not need permission to reproduce a brief quotation from a published document such as a magazine interview. But words spoken aloud in the workplace and words written for limited circulation (such as internal memos) are not within the public domain.

This principle also applies when research is done after the fact. For example, a researcher might want to analyze e-mail content between subject matter experts and technical communicators. From a researcher's perspective, the employees own the words in their e-mails. This approach does not contradict the fact that the company owns all intellectual property created by employees, including their words (as technical communicators, we know this only too well). But in respecting the rights and privacy of research participants, the researcher respects the right of individuals to control how their words and expressions are used for research, including whether they are used at all. So in this example, the researcher would need to get the informed consent of the e-mail authors.

On the other hand, if all the researcher wanted to do was to analyze the frequency of e-mail exchanges between subject matter experts and technical communicators during different phases of a product's life cycle, informed consent would not be an issue because employee words and identities would not be used as data.

Is It Research or Normal Evaluation?

Separating research from routine evaluation can sometimes be tricky. Although evaluation can be a type of research, not all evaluations are research. Pilot studies of new training programs or beta tests of new products are common examples of routine evaluations that do not, strictly speaking, constitute research. The guiding question should be, "If I were not doing research, would I have created this program or would I be evaluating it?"

For example, a company might introduce a new intranet information portal in a pilot study in one plant while using the existing documentation repository in other plants. The company's intent might be to compare productivity and error rates of employees who have access to the new portal to the productivity and error rates of employees using the existing format. Would informed consent be required? No. For one thing, it is well within the company's normal scope of operations to issue documentation and to require its use, and the new intranet portal surely fits that situation. The fact that the company is evaluating the portal's effectiveness before implementing it across the entire enterprise is just sound operating practice.

Getting Informed Consent

Informed consent can be obtained by having participants sign a form that describes the research and asserts their willingness to participate. The form should include a statement to the effect that the participant may choose to stop participating at any time.

Informed consent can also be made implicit in the act of participation. For example, an online survey can provide all the information that the form would have and stipulate that by participating in the survey the participant consents. This is one way to gain consent while still maintaining participant confidentiality.

Institutional Review Boards (IRBs)

Organizations that routinely conduct research involving human subjects will have IRBs that must approve any research. These boards are common in universities, research hospitals, and government health organizations such as the Centers for Disease Control (CDC). An IRB has its own format and procedures that you will have to follow if you are subject to its review. If you are a student researcher, your professor will tell you whether IRB approval is required.

Defining the Goal and the Research Questions

Now that you understand the components of the research report and the phases in the research process to conduct the research that generates that report, let us focus on getting started. Do not underestimate the importance of defining the research goal and questions; they are the steering force of your project. The earlier you can identify these key elements, the better off you are. In a long research program, such as a doctoral or a master's program, having your topic and goal defined early in your program is especially helpful; that way, many papers you write for individual classes can be geared toward your final research project. In effect, you turn as many classes as possible into an opportunity to add to your literature review and refine your research questions.

As daunting as the task might seem at first, it is quite manageable if broken down into steps. Table 2.1 shows a worksheet for approaching the task. Later pages in this chapter show filled-out samples of the worksheet.

Table 2.1 Worksheet for developing goal and research questions

Topic	Goal	Questions	Type	Methods

The first step is to identify the topic or topics that you are interested in. There is an exercise in Chapter 1 that got you thinking about potential research topics in technical communication. Now you must start to identify specific topics that interest you. Actually, it is not as hard as it might appear; you are probably sitting in the middle of dozens of research opportunities. To get started, there are two primary sources of research ideas: the literature and real-life problems.

Getting Research Ideas from the Literature

As you read research articles and books about technical communication, it seems that every topic has already been researched and there are no new topics. Do not dismiss a topic just because someone else has already researched it or it has been discussed at length in the literature. Most research is derived from research that has gone before it. Ironically, it is easier to get sponsorship around ideas that have already been researched than for a completely original idea. Academe and professional communities prefer gradual advances that build on and preserve conventional practices and knowledge over paradigm shifts that overturn what has gone before (Kuhn 1962).

So do not be discouraged if someone else has already done research and written on a topic you thought you had discovered. Instead, use it as an opportunity to seed your own research topic or agenda. Use the following checklist of questions to see whether existing research can fuel additional research on a particular topic:

- Can the research be updated? For example, if a study on online help was done 10 years ago, a legitimate question is, "Do its results still apply to readers today who might be more sophisticated users of online documentation?"

- Can the research be narrowed? For example, a study of Internet applications in general might conclude that programmatic techniques such as AJAX make user interactions richer and enhance the user experience. This conclusion could spawn a new study, however, on the effects that these techniques have on disabled users.
- Can the research be broadened? For example, a lot of research uses college students as the test sample because this is a convenient population for academic researchers to tap into. Therefore, a good topic for a research project could be to apply research that has been done on this narrow segment of the user population to a broader population, such as working adults.
- Can the research be challenged? Just because a researcher says something is true does not mean that it is. If a researcher's claim seems suspect to you, analyze what flaw in the researcher's assumptions or methodology might have led to the claim and design a research project to examine the potential flaws. For example, a usability study might conclude that users readily go to help when they encounter difficulty. Upon examination of the protocol, you see that the test respondents were told that this was a test of online help. You might challenge that this information predisposed them to use help and you could do a similar research project where users were not given this information.

Getting Research Ideas from Real-Life Problems

If we go back to our definition in Chapter 1 that research should inform decisions about technical communication practices, then any time a technical communicator is making a design or production decision, an opportunity for meaningful research could present itself. For example, if during a design meeting a proposal is made that users would benefit more from field-level help rather than screen-level help, this presents an opportunity to do research. Basically, questions like "Which way is better?" or "Which would users prefer?" or "How do users ...?" are all opportunities to get real data about real users rather than to speculate or wage abstraction wars in a conference room.

Using the Worksheet

Let us walk through an example showing how a student could identify a goal and research questions for a course-based research project. The first step is to identify at a high level some topics of interest. In our example, the student might have an interest in help files because that is what she works with at her job, and she might also have an interest in usability because she has been reading about it in the professional newsletter she gets every month. Table 2.2 shows these topics entered into the worksheet.

Table 2.2 Worksheet with topics identified

AU: Not cited in text.

Topic	Goal	Type	Questions	Methods
Help				
Usability				

The next step is to narrow down each topic to a specific goal. At this point it might be easier to think about the problem rather than the goal. For example, the student might think, "One of the big problems I face with help is the fact that I hear that our users don't use it." This, in turn, could spawn the goal statement "Understand why users are reluctant to use help."

Similarly, the student could also be involved in a debate at work over whether help should be presented in a traditional help file or as embedded help displayed as a part of the application interface itself. This could lead to the issue of which type of help is more usable. So, a goal for her second topic could be "To compare the usability of conventional help and embedded help." Table 2.3 shows the goals added to the worksheet.

Table 2.3 Goal statements defined

Topic	Goal	Questions	Type	Methods
Help	Understand why users are reluctant to use help			
Usability	Compare the usability of conventional help and embedded help			

At this point in the process, the student might start the literature review to understand the goal areas a little better. For her first topic, she might read articles that speculate that users stay away from help because they have gone there in the past and did not find it helpful. She might ask, "I wonder whether that is true." She might also have attended a workshop on writing online help that emphasized the importance of matching the help's content structure to the user's mental model of the task, and she wonders whether this could be an influencer. So, she now narrows the scope of her goal by formulating the following research questions:

1. What prior experiences with help do users have?
2. Does the structure of the help match the user model of the task?

She goes through a slightly different process with the next goal of comparing the usability of conventional help with embedded help. Here, she is driven more by the question "What does 'more usable' mean?" She reads some articles on usability and decides that frequency of use and time to get information would be good indicators of usability. This decision forms her research questions for that topic:

1. Do users use one type of help more than the other?
2. Do users find answers faster with one type than with the other?

Table 2.4 shows those questions added to the worksheet.

Before choosing between one topic or the other, the student should identify the type of research each would be and describe what methods she or he would use. This step will help the student gauge the scope of the project. See Table 2.5.

Now the student can make an informed choice about which project she would rather undertake for her course assignment. Some students spend a lot of time dithering about this decision and, quite frankly, lose a lot of valuable time that would be better

Table 2.4 Goals and questions

Topic	Goal	Questions	Type	Methods
Help	Understand why users are reluctant to use help	1. What prior experiences with help do users have? 2. Does the structure of the help match the user model of the task?		
Usability	Compare the usability of conventional help and embedded help	1. Do users use one type of help more than the other? 2. Do users find answers faster with one type than with the other?		

Table 2.5 Completed worksheet

Topic	Goal	Questions	Type	Methods
Help	Understand why users are reluctant to use help	1. What prior experiences with help do users have? 2. Does the structure of the help match the user model of the task?	Interpretivist	Interviews Field observations
Usability	Compare the usability of conventional help and embedded help	1. Do users use one type of help more than the other? 2. Do users find answers faster with one type than with the other?	Empirical	Usability test Comparisons of average frequencies and average times

spent doing research. The advice we have is, "This is your first research project, not your last." In other words, you can do both eventually, but you cannot do both right now. Pick one to do now and *get started.*

Exercise 2.2: Identifying research goals and questions

Use the worksheet introduced in Table 2.1 to do this exercise.

1. Identify at least three topics of interest.
2. For each topic, write a tentative goal statement.
3. For each goal, write two to five research questions.
4. Based on the questions, for each goal classify the type of research you would do and list what methods you would probably employ.
5. Based upon your skills, access to potential data, and time available for the research, pick one of the goals for your research project. ■

There! The hardest part is over. The remaining sections of the book will help you with the rest.

Answer Key

Exercise 2.1

- Statement of the Research Problem ("Introduction," 130–132)
- Review of the Literature ("Introduction," 130–132, specifically, the first three paragraphs)
- Description of the Methodology ("Experimental Methods," 132–137)
- Analysis of the Data ("Results and Discussion," 137–143)
- Conclusions ("Conclusions," 143–145)

Exercise 2.2

- The answers to this exercise will be unique to each person who prepares it, so there is no key to this exercise.

References

Charles, C.M. 1998. *Introduction to Educational Research*, 3rd ed. New York: Addison-Wesley Longman.

Kuhn, T. 1962. *The Structure of Scientific Revolutions.* Chicago, IL: University of Chicago Press.

Office of Human Subjects Research. 1979. The Belmont report: Ethical principles and guidelines for the protection of human subjects of research. http://ohsr.od.nih.gov/guidelines/belmont.html.

Smudde, P. 1991. A practical model of the document-development process. *Technical Communication* 38:316–323.

The Nuremberg Code. 1949. *Trials of War Criminals Before the Nuremberg Military Tribunals Under Control Council Law No. 10*, vol. 2. Washington, D.C.: U.S. Government Printing Office.

3

Reviewing the Literature

Introduction

We begin this chapter by exploring the differences between primary and secondary research. We then consider the reasons for doing secondary research as part of a primary research project and explore the best way to conduct a review of existing literature on a topic. We examine reference lists and reference citations prepared according to both the *Publication Manual of the American Psychological Association* (American Psychological Association 2001) and *The Chicago Manual of Style* (University of Chicago 2003), and we describe the process for preparing an annotated bibliography of sources. Finally, we discuss the process of writing a literature review.

Learning Objectives

After you have read this chapter, you should be able to

- Differentiate between primary and secondary research
- Describe the purposes of the literature review:
  - Determining what has already been learned about the area to be researched
  - Identifying gaps in the research
  - Educating the reader
  - Establishing credibility with the audience
- Conduct a review of online and library sources and assemble an initial reading list
- Cite references and build a reference list in accordance with the *Chicago Manual of Style* and the *Publication Manual of the American Psychological Association*
- Describe and evaluate the content of your sources
- Prepare an annotated bibliography
- Write a literature review

Primary and Secondary Research

The major focus of this book is primary research. Our goal is to help you learn how to perform—and how to be an informed consumer of—primary research. A literature review, the focus of this chapter, is an example of what is usually called secondary research.

You may have heard your professors distinguish between primary or "original" research, and secondary or "derivative" research. Although the labels *primary* and *secondary* are helpful distinctions, the terms *original* and *derivative* are less helpful and can even be misleading in some respects.

Primary research involves formulating and testing a hypothesis, collecting information through observation, or conducting a survey as described to Chapter 6 of this text, and then reporting the results to others in the field. Such research is called *primary* because it adds to the existing body of knowledge in the field. For example, a researcher might ask a team of technical communicators from the United States and another from China to separately write, design the layout, and prepare illustrations for a set of instructions for a bilingual e-commerce Web site intended for an audience of beginner Internet users in their countries, and then have each set of instructions (including any text in the illustrations) professionally translated to the other language. Once the translations have been completed, the researcher might design and conduct a series of usability tests to explore how the design and the rhetorical patterns of each set of instructions affected the users of the other culture in terms of their speed in completing various procedures. This is an example of primary research because it would add to the current state of knowledge about the relationships among cultural expectations, rhetorical patterns, and page design.

Secondary research is conducted by reading reports of previous research, analyzing the results, and then formulating a synthesis of the current state of knowledge on a topic. This research is called *secondary* because it draws on reports of previous primary and secondary research. For example, a researcher interested in intercultural communication would want to know the results of all primary research that has investigated the effects of the rhetorical patterns and design of one culture on users from a different culture, as well as any previous secondary work on the topic. The result would be an example of secondary research because it would describe the research that has already been done on the topic.

Although the distinction between primary research and secondary research is an important one, the terms *original* and *derivative* are not entirely accurate in describing these two classifications of research because primary research can be derivative and secondary research can make very significant contributions to a field.

How can primary research be derivative? Well, one of the most important characteristics of primary research is that it should be repeatable and that when repeated, it should produce the same results each time. When we formulate and test a hypothesis experimentally, when we collect information through observation, and when we conduct a survey, we draw conclusions or make generalizations based on our analysis of the data that we collect. But all primary research is based on a sampling of the total population—a relatively small number of potential users of an MP3 player, for example, or every twentieth registered user of a software product. And sometimes the samples that researchers test, observe, or survey are less than ideal—we may draw on students as samples of convenience, rather than people from a wider range of ages and educational backgrounds, because students are easier to recruit and study, especially for researchers on university faculties.

Because the samples used in much primary research are often relatively small and may be less than ideally representative of the total population, the conclusions based on that research are usually stated tentatively. In their conclusions, the authors of such studies typically encourage other researchers to investigate the same questions using different sample populations, and other researchers sometimes duplicate the studies of others in this way by using a larger or more representative sample, or by varying the sample in some way. Such primary research that attempts to duplicate or test the reliability of research that others have performed is thus, in a sense, derivative, although it is quite important because it verifies the generalizability of results obtained by earlier researchers.

Similarly, secondary research can be quite significant despite the fact that it draws on the primary and secondary research performed by others. For example, an annotated bibliography of primary and secondary sources on intercultural communication could make a very important contribution to the field of technical communication by summarizing earlier work that has been done on this topic. Similarly, a bibliographical essay could be an essential source for other students of intercultural communication because of the synthesis it provides and the insights it provokes.

The Purposes of a Literature Review

Discovering What Has Already Been Done

One of the most important reasons for reviewing the literature is the heuristic function of this task: Discovering what others have already done in your area of interest will give you insights into aspects that you want to investigate and suggest contributions that you can make to the field.

For example, suppose you are interested in doing research on intercultural communication. Conducting a thorough review of existing work in the field will reveal that many researchers have drawn on the concept of cultural dimensions described by Geert Hofstede (1994, 2001) to formulate their research studies—the effect of cultural differences in uncertainty avoidance on the willingness to read software documentation, or the effect of cultural differences in power distance on willingness to follow directions precisely. Reading everything you can find on the theoretical contributions of Hofstede and the research studies based on his theory of cultural dimensions could be very influential in helping you zero in on a primary research study you would like to conduct.

Whether you intend to prepare an extended survey of the existing literature or not, you need to know what research has already been done on your topic so you do not needlessly repeat work that has already been done. For example, suppose that you are interested in the intersection between page design and reading comprehension. By reviewing the existing literature on the topic, you will likely conclude that there is not much point in studying the relationship between printed line length and user comprehension of the information provided because a lot of work has already been done on that subject with quite similar results. Even if you are not working on a PhD dissertation that is supposed to make an original contribution to knowledge on your topic, you probably do not want to go over ground that others have already covered quite adequately.

Identifying Gaps in Existing Research

Knowing what research has already been done on your topic of interest also helps you discover what remains to be studied. What are the gaps in existing studies that still need to be filled in? What are the areas that have not been explored at all? For example, suppose that you have read all the work that has been previously done on the relationship between the length of printed lines and the comprehension of users. You believe initially that this topic is tapped out—there is nothing useful remaining to be done. And then perhaps you realize that all the existing studies of this subject have involved texts in Western languages and have used North American and European users. You wonder whether you would get the same results if you studied text in Chinese with a group of Chinese users.

Similarly, reading the research about printed line length and comprehension might cause you to wonder whether there is a relationship between the spatial distance between a procedural step and the graphic that illustrates it, and the user's comprehension of the instruction. This research question might not have occurred to you if you had not read this earlier research.

Educating Readers

So far, we have talked about the illuminating quality of a literature review for the researcher, but the researcher's audience will also benefit from this effect because most research reports include a summary of related research on the topic. This literature review within a book, article, or paper serves several functions.

First, the literature review situates the current study. It provides background information that explains why you decided to do your research and often describes how the previous work on the topic helped you frame your research question. This background is helpful to your readers because it provides a kind of intellectual history that explains how you became interested in pursuing it and how you have approached it.

Second, the literature review explains the context of your research to those who read your report. Many—perhaps most—will not have read all of the previous work that you have reviewed before designing and conducting your experiment, observation, or survey. Your review of the literature provides your own readers with the essential information they need to know about your area of inquiry.

Establishing Credibility

One important effect of the review of the literature included in your article or paper is that it establishes your credibility with your audience, for both those who are familiar with the work you discuss and for those for whom it is new. In both cases, you demonstrate your awareness of the work that others have done. Your subsequent description of your research methodology, presentation and discussion of the data you collected, and conclusions about your research question thus gain prestige in your readers' minds because of the relationship—the intellectual pedigree—you have established between the previous work and your own.

On the other hand, imagine the effect in a reader's mind if your article or paper does not mention an important prior work on the topic, especially if your research contradicts the findings or observations of the earlier researcher. That reader is unlikely to give your conclusions much intellectual weight because it will seem that you have failed to do your job of thoroughly surveying the previous significant work on the subject.

Conducting a Literature Review

Conducting a literature review is actually rather simple. You begin by assembling as comprehensive a list of sources on your general area of study as possible, and then refine that list to include only the works that deal with your topic in some detail. The following tools will make your task easier.

- *St. Martin's Bibliography of Business and Technical Communication*: Although not a comprehensive bibliography of the field, this annotated and classified listing of 376 books and articles compiled by Gerald Alred and published in 1997 is definitely a groundbreaking work. Do not limit your research to Alred's selection of work published before 1997, but this is probably the first place you should go to find books and articles on your topic.
- Library catalogs and databases: If you have access to a good college, university, or public library, you can search for books in its "card" catalog (now typically online) and for periodical article citations in the databases to which the library subscribes. Two databases are particularly helpful because they also provide the full text of the articles they index:
  - EBSCOhost Communication and Mass Media Complete
  - ProQuest Research Library

  When you find articles in a database, note any keywords different from those you used in your search, and search for additional articles using those keywords.
- Annual ATTW bibliography: The Association of Teachers of Technical Writing (ATTW) publishes an annual bibliography of books, articles, review essays, and book reviews. The bibliography for a year typically appears in the autumn issue of *Technical Communication Quarterly* for the following year. This journal changed publishers in 2004, and the publication of the 2003 ATTW annual bibliography did not appear until autumn 2005.
- EServer Technical Communication Library: EServer (http://tc.eserver.org/) contains an annotated catalog of more than 10,000 articles and other materials of importance to the field. It is searchable and provides links to most of the sources included in the database.
- "Recent and relevant": The "Recent and relevant" department of *Technical Communication* provides abstracts of articles appearing in the other journals in the field as well as journals in many related fields. The abstracts typically appear about six months following publication, but there are occasionally lags and lack of coverage.
- Book reviews in journals: Reviews published in *IEEE Transactions on Professional Communication*, *Journal of Business and Technical Communication*, *Journal of Technical Writing and Communication*, *Technical Communication*, and *Technical Communication Quarterly* cover virtually all titles in the field of technical communication. *Technical Communication* publishes reviews of books in a variety of cognate fields as well. Be aware that reviews may not appear for two years or more after publication, so scanning these reviews is not a substitute for other ways to locate recent book-length works in the field, but reviews are typically more timely than many other sources.
- Online booksellers' catalogs: Amazon, Barnes & Noble, and other online booksellers allow you to search their databases by topic and provide full bibliographic citations of works contained there, as well as the ability to order the books online.
- Works cited by other sources: Search through the reference list or bibliography in each of the sources you have located. What additional works do those sources refer to? They may prove helpful to you.

Creating a List of References

You probably started writing research papers in high school, if not earlier, and you (and your teachers) certainly hated one aspect of such assignments more than any other: Learning (and teaching) whatever citation style was prescribed by the handbook you used in the course. These details can be a pain because they seem so trivial and arbitrary.

Why do we bother with footnotes, endnotes, or in-text citations? Why the hassle with bibliographies or lists of references? The only reason is to help those who want to further explore the work we have done—perhaps to reproduce our primary research or to use our secondary research as a jumping-off point for their own study. In other words, we compile citations of sources in standard formats to contribute to the advancement of our field of study.

The most common forms of references and in-text citations used in technical and professional communication today are what are commonly called APA and Chicago styles.

APA Style

The *Publication Manual of the American Psychological Association* has its roots in a seven-page article in the APA's *Psychological Bulletin* in 1929. Now in its fifth edition and more than 400 pages long, the APA manual is used by more than 1,000 journal and book publishers, as well as in academic papers in the behavioral and social sciences. *Technical Communication Quarterly* uses APA style for in-text citations and lists of references.

APA maintains a Web site that provides updates to the *Manual* at http://www.apastyle.org/.

The following are examples of the more common types of reference citations provided in APA style. If you are not able to find an example that corresponds to a work you need to cite, consult the APA manual.

Book with one author

Rubin, J. (1994). *Handbook of usability testing: How to plan, design, and conduct effective tests*. New York: Wiley.

Book with two to six authors

Stone, D., Jarrett, C., Woodroffe, M., & Minocha, S. (2005). *User interface design and evaluation*. San Francisco: Morgan, Kaufmann.

Note: If a book has more than six authors, use "et al." after the comma following name of the sixth author, without using an ampersand (&).

Editor, translator, or compiler

Sides, C. (Ed.). (2006). *Freedom of information in a post 9-11 world*. Amityville, NY: Baywood.

Essay in an edited collection

Bosley, D. (2002). Jumping off the ivory tower: Changing the academic perspective. In B. Mirel & R. Spilka (Eds.), *Reshaping technical communication: New directions and challenges for the 21st century* (pp. 27–39). Mahwah, NJ: Erlbaum.

Chapter or portion of an edited volume

Aristotle. (1984). The rhetoric, Book I. In W. R. Roberts & I. Bywater (Trans.), E. P. J. Corbett (Intro.), *The rhetoric and the poetics of Aristotle* (pp. 19–89). New York: Random House.

Preface, foreword, introduction, and similar parts of a book

Redish, J. C. (2002). Foreword to *reshaping technical communication,* B. Mirel and R. Spilka (Eds.), (pp. vii–xii). Mahwah, NJ: Erlbaum.

Electronic book

Clark, J. (2002). *Building accessible Websites.* Retrieved March 8, 2006, from http://joeclark.org/book/

Article in journal in which the entire volume is paginated consecutively

Giammona, B. (2004). The future of technical communication: How innovation, technology, information management, and other forces are shaping the future of the profession. *Technical Communication, 51,* 349–366.

Article in journal in which each issue is paginated separately

Marcus, A. R., & Gould, E. W. (2000). Crosscurrents: Cultural dimensions and global Web user-interface design. *Interactions, 7(4),* 32–36.

Article in an electronic journal

Meeks, M. G. (2004). Wireless laptop classrooms: Sketching social and material spaces. *Kairos 9.* Retrieved March 8, 2006, from http://english.ttu.edu/kairos/9.1/binder2.html?coverweb/meeks/index.html

Magazine article

Hart, G. J. S. (2000, March). The style guide is dead: Long live the dynamic style guide. *Intercom,* 12–17.

Newspaper article

Schwartz, J. (2003, September 28). The level of discourse continues to slide. *New York Times,* p. D-12.

Book review

Farkas, D. K. (2005). Review of the book *Beyond bullet points: Using Microsoft PowerPoint to create presentations that inform, motivate, and inspire. Technical Communication, 52,* 465–467.

Thesis or dissertation

Dayton, D. (2001). *Electronic editing in technical communication: Practices, attitudes, and impacts.* Unpublished doctoral dissertation, Texas Tech University.

Unpublished presentation at a meeting or conference

Note: A paper included in the published proceedings of a meeting is treated as an essay in an edited collection.

Doumont, J.-l. (2005, July). *Effective slides: Design, construction, and use.* Workshop presented at the International Professional Communication Conference, Limerick, Ireland.

Personal communications

Note: A personal communication such as a telephone conversation, letter, or e-mail message is not included in the list of references. Instead, it is cited in the text, as follows:

David Dayton observed that . . . (personal communication, October 31, 2002).

An entire Web site

Access Board. *Section 508.* Retrieved March 8, 2006, from http://www.section508.gov

An article, document, or short work from a Web site

Access Board. Section 508 standards. *Section 508.* Retrieved March 8, 2006, from http://www.section508.gov/index.cfm?FuseAction=Content&ID=12

Chicago Style

The Chicago Manual of Style—often referred to as *CMOS*—is currently in its 15th edition and is nearly 1,000 pages long. It originated in the 1890s as a single page of instructions for proofreaders at the University of Chicago Press, and even today it reflects the press's house style. Its comprehensive guidance for preparing manuscripts and its detailed descriptions of the publication process have established it as the style guide of choice of many journal and book publishers.

CMOS actually offers two styles of references:

- A style using footnotes or endnotes and a bibliography of sources at the end of the work, commonly used by writers in the humanities
- A style using author-date citations in the text and a list of references at the end of the work, commonly used by writers in the natural and social sciences

In this book, we will address the author-date Chicago style, which *Technical Communication* uses for in-text citations and lists of references.

The University of Chicago Press has created a Web site for *CMOS* at http://www.chicagomanualofstyle.org/cmosfaq.html where you can find helpful frequently asked questions (FAQs) and sample Chicago-style references.

The following are examples of the more common types of reference citations provided in Chicago author-date style. If you are not able to find an example that corresponds to a work you need to cite, consult *The Chicago Manual of Style.*

Book with one author

Rubin, Jeffrey. 1994. *Handbook of Usability Testing: How to Plan, Design, and Conduct Effective Tests.* New York: John Wiley & Sons.

Book with two or three authors

Hackos, JoAnn T., and Janice C. Redish. 1999. *User and Task Analysis for Interface Design.* New York: John Wiley & Sons.

Book with more than three authors

Stone, Debbie, Caroline Jarrett, Mark Woodroffe, and Shaailey Minocha. 2005. *User Interface Design and Evaluation.* San Francisco: Morgan, Kaufmann.

Editor, translator, or compiler

Sides, Charles, ed. 2006. *Freedom of Information in a Post 9-11 World.* Amityville, NY: Baywood Publishing Co., Inc.

Essay in an edited collection

Bosley, Deborah. 2002. Jumping off the ivory tower: Changing the academic perspective. In *Reshaping Technical Communication: New Directions and Challenges for the 21st Century*, ed. Barbara Mirel and Rachel Spilka, 27–39. Mahwah, NJ: Lawrence Erlbaum Associates.

Chapter or portion of an edited volume

Aristotle. 1984. The rhetoric, Book I. In *The Rhetoric and the Poetics of Aristotle*, trans. by W. Rhys Roberts and Ingram Bywater, intro. by Edward P. J. Corbett, 19–89. New York: Random House.

Preface, foreword, introduction, and similar parts of a book

Redish, Janice C. 2002. Foreword to *Reshaping Technical Communication*, ed. Barbara Mirel and Rachel Spilka, vii–xii. Mahwah, NJ: Lawrence Erlbaum Associates.

Electronic book

Clark, Joe. 2002. *Building accessible Websites.* http://joeclark.org/book/.

Article in journal in which the entire volume is paginated consecutively

Giammona, Barbara. 2004. The future of technical communication: How innovation, technology, information management, and other forces are shaping the future of the profession. *Technical Communication* 51:349–366.

Article in journal in which each issue is paginated separately

Marcus, Aaron R., and Emilie West Gould. 2000. Crosscurrents: Cultural dimensions and global Web user-interface design. *Interactions* 7 (4): 32–36.

Article in an electronic journal

Meeks, Melissa Graham. 2004. Wireless laptop classrooms: Sketching social and material spaces. *Kairos.* http://english.ttu.edu/kairos/9.1/binder2.html?coverweb/meeks/index.html.

Magazine article

Hart, Geoffrey J. S. 2000. The style guide is dead: Long live the dynamic style guide. *Intercom* (March): 12–17.

Newspaper article

Schwartz, John. 2003. The level of discourse continues to slide. *New York Times*, September 28, sec. 4.

Book review

Farkas, David K. 2005. Review of *Beyond Bullet Points: Using Microsoft PowerPoint to Create Presentations That Inform, Motivate, and Inspire,* by Cliff Atkinson. *Technical Communication* 52:465–467.

Thesis or dissertation

Dayton, David. 2001. Electronic editing in technical communication: Practices, attitudes, and impacts. PhD diss., Texas Tech Univ.

Unpublished presentation at a meeting or conference

Note: A paper included in the published proceedings of a meeting is treated as an essay in an edited collection.

Doumont, Jean-luc. 2005. Effective slides: Design, construction, and use. Presentation at the International Professional Communication Conference, July 10–13, in Limerick, Ireland.

Personal communications

Note: A personal communication such as a telephone conversation, letter, or e-mail message is not included in the list of references. Instead, it is cited in the text, as follows:

In an e-mail message to the author on October 31, 2002, David Dayton observed that . . .

An entire Web site

Access Board. *Section 508*. http://www.section 508.gov.

An article, document, or short work from a Web site

Access Board. Section 508 standards. *Section 508*. http://www.section508.gov/index.cfm?FuseAction=Content&ID=12.

Notable Differences Between APA and Chicago Reference Styles

As you have probably noticed, there are many similarities between the APA and Chicago styles for reference list entries, but there are also several differences. Let us consider the most significant of those differences in an effort to better understand the two systems.

- **Authors' names:** APA reverses all author names—surname first for multiple author works, uses only the initials of first and middle names, and uses an ampersand (&) before the last author's surname for multiple author works. Chicago reverses only the first author's name, uses full first names and middle initials, and uses *and* (not the ampersand) before the last author's name.
- **Publication dates:** APA places parentheses around the publication date and follows the parenthetical date with a period; Chicago does not use parentheses.
- **Publisher names:** APA shortens publisher names; Chicago uses full publisher names.
- **Journal titles:** APA follows the title with a comma. Chicago does not use a comma after the journal title.
- **Journal volumes/issue numbers:** APA italicizes the volume and issue numbers of journals; Chicago does not.

- **Newspaper articles:** APA provides section and page references for newspaper articles; Chicago omits this information.
- **URLs:** APA provides retrieval dates for URLs, and does not use a period at the end of the URL. Chicago does not provide retrieval dates and follows the URL with a period.

Exercise 3.1: Preparing reference citations

Using Chicago or APA style (choose the form specified by the publisher, journal, or course for which you are preparing a submission), assemble a reference list that includes the appropriate citations for the following works.

1. Dissertation for the doctor of philosophy degree written by Kim Sydow Campbell at Louisiana State University in 1990. The title is "Theoretical and Pedagogical Applications of Discourse Analysis to Professional Writing."
2. Article on pages 21 through 23 in the January 2001 issue of *Intercom* magazine entitled "Technical Communicators: Designing the User Experience." The author is Lori Fisher.
3. E-mail message from Robin Willis to you dated March 8, 2006.
4. Web page (URL: http://www.useit.com/papers/webwriting/writing.html) entitled "Concise, SCANNABLE, and Objective: How to Write for the Web." The authors are John Morkes and Jakob Nielsen.
5. Conference paper entitled "Interact to Produce Better Technical Communicators: Academia and Industry" published in 1995. It was written by Thea Teich, Janice C. Redish, and Kenneth T. Rainey, and appeared on pages 57–60 in the *Proceedings of the STC 42nd Annual Conference*. The publisher is the Society for Technical Communication, located in Arlington, Virginia.
6. Article in the *New York Times* by Eduardo Porter. The title is "Send Jobs to India? Some Find It's Not Always Best." It was published on April 28, 2004, and it appeared in section A on page 1.
7. Usability Body of Knowledge Web site of the Usability Professionals' Association (URL: http://www.usabilitybok.org/).
8. Book entitled *Designing Visual Language: Strategies for Professional Communicators*. It was published by Allyn and Bacon, which is located in Needham Heights, Massachusetts. It was written by Charles Kostelnick and David D. Roberts and published in 1998.
9. Article (URL: http://english.ttu.edu/kairos/7.3/binder2.html?coverweb/fishman/index.html) in the online journal *Kairos* by T. Fishman. The title is "As It Was in the Beginning: Distance Education and Technology Past, Present, and Future," and it was published in 2002.
10. Article in the third issue of *Technical Communication*'s volume 47 in 2000. The author is Judith Ramey. The title is "Guidelines for Web Data Collection: Understanding and Interacting with Your Users." It appeared on pages 397 through 410.

■

Describing and Evaluating Your Sources

Whether you are planning a formal annotated bibliography as a secondary research project or you are surveying the relevant literature as an initial step in a primary research project, you will find that it is helpful to prepare annotations of your sources for several reasons:

- Composing descriptive annotations will help you better understand—and remember—the content of individual books, articles, and other sources about your topic.
- Preparing descriptive annotations will also help you see connections among the various separate works on your research topic.
- Writing evaluative annotations will help you explicitly gauge the relative importance of the existing work in your area of interest.
- Creating evaluative annotations will also help you see the strengths and weaknesses of the existing work, and potentially draw on the strengths and avoid the weaknesses in your own research.

An annotated bibliography consists of a series of entries, each corresponding to a single work you have selected for inclusion. The entries are typically arranged alphabetically by the last names of the authors (by the name of the first author in the case of multiauthor works). Long, annotated bibliographies (those containing more than 25 or 30 entries) are frequently classified by relevant subtopic first, with the entries for each subtopic then arranged alphabetically by the authors' last names.

Each entry consists of three parts:

1. A full bibliographic citation, using the preferred style (such as Chicago or APA) of the journal, publisher, or course for which you are preparing the bibliography
2. A description of the work's content
3. An evaluation of the work's significance

Because we have already talked at length about preparing bibliographic citations using APA and Chicago style, let us examine how to write the descriptive and evaluative annotations.

Writing Descriptive Annotations

Your descriptive annotation is essentially an abstract of the book, article, or other work. It may take either of two forms.

- **You may reproduce the author's abstract** if one exists, but you must place it between quotation marks and include the page numbers or URL (if it is published only online and is not included as a part of the book, article, or other source itself). For example:

> "This article looks at the future of technical communication from the point of view of many of its most seasoned and influential practitioners. And it wraps that point of view around the themes of the [New York Polytechnic University management of technology master's] program—innovation, global concerns, managing technical leaders and practitioners, the impact of new technologies, and the future

role of technologists in organizations. It concludes by providing a series of recommendations for the future direction of the profession." http://www.ingentaconnect.com/content/stc/tc/2004/00000051/00000003/art00002.

Although using the published abstract makes preparing the descriptive part of the annotation easier, preparing your own abstract will help you better understand the content of the book, article, or other source. Furthermore, your own abstract may be superior to that provided by the work's author. (You would be surprised how poor some published abstracts are!)

- **You may prepare your own descriptive abstract**. A good abstract provides readers with all of the essential information they need about the book, article, or other source, including the research question, research method, research results, analysis of the results, and conclusions. Writing your own abstract will help you master the intricacies of the work and better understand its significance, thus helping you write the evaluative part of the annotation. For example:

Giammona examines the future of the profession based on interviews with and survey responses from experienced practitioners and academics. She organizes her discussion using the following themes: our future role in organizations, management concerns, our contributions to innovation, global concerns, education of future professionals, and relevant technologies. She also provides recommendations of future directions that she believes will help the profession survive and thrive in the future.

Include only essential information in the abstract. For example, if the author reports that the difference between the experimental and control groups' performance is statistically significant, that information is important and should be included in the descriptive annotation. The tests that the author has used to determine the statistical significance are not essential.

Descriptive annotations are typically 100 to 250 words long, depending on the length and complexity of the work you are abstracting.

Writing Evaluative Annotations

Preparing evaluative annotations is a very challenging task, especially to those who are new to technical communication research in general or new to the particular area of study. You might ask yourself whether you are qualified to evaluate the work of others, especially if the authors are well-known experts in the field. In fact, as someone new to the area of inquiry or to research in the field generally, you are in a perfect position to gauge the importance of the works you are analyzing.

First, you will be reading the books, articles, and other sources over a relatively short period of time rather than over a period of years as the works appear. By reading and digesting all the work related to your topic of interest more or less concurrently, you will find it much easier to compare these sources than someone who has read them over a much longer span of time.

Second, if you read the works chronologically in the order they appeared, you will find it quite easy to determine which works represent truly original research and which are derivative. One way of doing this is to notice which earlier sources other authors cite, especially in identifying their research questions and selecting a research methodology.

Finally, take note of which books, articles, and other sources represent the results of primary research and which are based only on secondary research. How significant are the results of the primary works? How successful are the authors of the secondary works in formulating a helpful synthesis of the primary and secondary work that others have done?

Your evaluative annotations should draw comparisons, where appropriate, to other works you are examining. To ensure that your evaluative comments are accurate and helpful, you should revisit and revise each one after you have completed all your reading and have a first draft of the descriptive and evaluative annotations of all your sources.

Here is a sample evaluative annotation of Barbara Giammona's article described in the preceding text.

> This article is significant because it is based on the views of 28 leaders in the field. The many insights in this in-depth qualitative research study will help technical communication practitioners (especially managers) anticipate and respond to trends in the profession, and will help educators prepare students for rewarding, long-lasting careers.

Evaluative annotations are typically 50 to 100 words long, depending on the length and importance of the work you are assessing. It is usually helpful to separate the descriptive and evaluative components of your annotations, devoting a paragraph to each. Begin with the abstract, and conclude with the evaluation.

Exercise 3.2: Composing an annotated bibliography

Prepare an annotated bibliography of at least twelve sources on your research topic. Make sure that you include both books and periodicals, both print and online sources. Prepare the citations using either APA or Chicago style, based on the preference of the journal, publisher, or course for which you are preparing the bibliography. Ensure that your annotations include both a descriptive and an evaluative component. ■

Writing The Literature Review

Annotated Bibliography vs. Literature Review

Now that you have completed an annotated bibliography on your topic, let us consider the differences between that genre and the literature review, as well as the reasons why you would choose one over the other.

As we have seen, an annotated bibliography is a listing of existing primary and secondary works on a topic. Each entry in the bibliography consists of a full bibliographic citation and both descriptive and evaluative comments about the book, article, or other work. The bibliography is usually arranged alphabetically by author, and bibliographies that include a large number of sources are often classified by subtopic. Many researchers prepare annotated bibliographies for their own use as a preliminary step in conducting their own primary or secondary research on the topic. They are also frequently assigned in courses to help students learn secondary research techniques or prepare them to explore a topic further in a primary research project. Other annotated bibliographies are published as articles, sections of books, or parts of Web sites addressing the topic.

A literature review is always prepared for an audience other than the researcher. It is typically a section of a research article or paper that presents a synthesis of the most important primary and secondary research relevant to the primary research project the article or paper is reporting on. Rather than being organized by source like an annotated bibliography, the literature review is organized by topic. For example, the literature review for a usability study of an e-commerce Web site might include a section on the usability methodology to be used, a section on findings of previous usability studies of e-commerce applications, and another section on the user populations to be sampled. The article or paper of which the literature review is a part ends with a list of references that provides a complete citation of each work mentioned in the literature review.

You will find that an annotated bibliography is easier and quicker to prepare because it does not require you to do the same amount of synthesizing of the sources that is necessary to write a literature review. You must, of course, read each work that you include in the annotated bibliography, and you must intellectually digest the sources to the extent necessary to describe them and evaluate their relative significance. But an annotated bibliography does not require the same degree of familiarity with the interrelationships of the sources that you must possess to produce a good literature review.

Preparing the Literature Review

To ensure the effectiveness of your literature review, you must define its purpose and audience, you must determine the most effective organization of the information, and you must decide the appropriate level of detail for the information.

Purpose and Audience. The audience of your literature review will be the readers of your article or paper. That audience might be the readers of the journal to which you intend submitting a manuscript reporting the results of your research. Or it may be the instructor and other students in a course. Whoever that audience is, analyze those potential readers as you would any other audience. How much are they likely to already know about your topic and previous research about it? What is their level of interest in the topic? How will the information you report in the literature review section of your paper or manuscript be of interest and use to them in their own research or practice?

In the context of a report on primary research that you have conducted, the purpose of your literature review is to provide a brief overview of work previously done on the topic. For example, if you are conducting primary research on the usability of a brochure aimed at a teen audience about alcohol using the plus-minus method, the literature you would include in your review would include the findings of previous usability studies about brochure design, teen audiences, and attitudes toward alcohol, as well as studies of the plus-minus method. Because it is intended to provide your audience with the broad strokes of previous research, your literature review would typically run to only a few pages and would not be as detailed as the annotated bibliography you have prepared for your own use. It would also probably include only the most important works, not a comprehensive treatment of every book and article you consulted while doing your secondary research.

Organization

There is no predetermined formula for organizing a literature review, but consider the following questions as you decide how best to arrange the results of your secondary research.

- What audience or user populations have previous researchers focused on?
- What methods or techniques have they used in conducting their research and analyzing the data they gathered?
- What are their major findings?
- What conclusions have they reached?
- Did they expect results that did not materialize or that were not conclusive?
- What work remains to be done?

Level of Detail

Remember that your purpose is to provide your readers with the information they need to understand what your research area is about, what work has been done already, and what remains to be explored. You may also need to include literature addressing methodology or analysis techniques if they are innovative or unusual. The literature review is not intended as a comprehensive or detailed account of all the work that has been done in an area before your project. Instead, it provides novices in the area the basics they need to understand your project and experienced researchers a recapitulation of what they have already read themselves.

Determining the appropriate level of detail for your literature review will depend in large part on its purpose, the number of sources you are surveying, and the amount of information those sources contain about your topic of interest.

If you are surveying the literature as part of a report of primary research, your literature review will likely form a relatively brief part of the larger paper or article. In this type of literature review, you will typically emphasize the highlights of the major works, whereas the less important works receive only a brief mention if they appear at all. If you are writing a stand-alone bibliographic essay, you will have a larger scope in which to work. As a result, you will probably provide a much greater amount of detailed information about the various sources you review. Compare the space that Peter MacKay devotes to his fourteen sources in "Establishing a corporate style guide: A bibliographic essay" (reprinted in Chapter 7) with the attention that Sam Dragga gives to more than twenty sources in the second section of "'Is this ethical?' A survey of opinion on principles and practices of document design" (reprinted in Chapter 10).

The number of sources and the space they devote to your specific topic are the other major influence on the level of detail you provide. In general terms, the more important—or the more general—your topic, the more sources you will find about it as you assemble a working bibliography, and the more information those sources will contain. If your topic is narrower, the task of reviewing the literature becomes more reasonable. For example, a very large number of sources deal with the role of the audience in technical communication, but the use of the second person in writing procedures would yield fewer sources.

Citing Your References in the Text

Any time you quote, paraphrase, or summarize information from any source, you must cite the source. Both APA and Chicago styles use parenthetical author-date citations rather than footnotes or endnotes. Here are some examples.

APA In-Text Citations

Simple author-date citation

Sans-serif type is more effective for online text because of its legibility (Schriver, 1997).

Citation with author's name incorporated into the text

Schriver (1997) recommends sans-serif type for online text because of its legibility.

Citation for a direct quotation

Choose sans-serif type for online text "because of its simple, highly legible, modern appearance …" (Schriver, 1997, p. 508).

Chicago In-Text Citations

Simple author-date citation

Sans-serif type is more effective for online text because of its legibility (Schriver, 1997).

Citation with author's name incorporated into the text

Schriver (1997) recommends sans-serif type for online text because of its legibility.

Citation for a direct quotation

Choose sans-serif type for online text "because of its simple, highly legible, modern appearance …" (Schriver, 1997, 508).

As with reference lists, you will notice that despite the general similarities, there are some differences between APA and Chicago styles. APA uses a comma between the author's name and the date, and uses *p.* or *pp.* before page numbers for citations of direct quotations.

As we noted earlier, you must provide a citation any time you summarize, quote, or paraphrase one of your sources. Using another person's intellectual property without attribution is known as plagiarism, and that is an offense you want to avoid at any cost!

How do you know when to cite a source? The easiest rule of thumb is to think back to the point before you began your secondary research on the topic. What did you know about the subject then? Any information that does not fall within your knowledge before beginning your research on the topic likely requires a citation.

An exception to this rule is what is called "common knowledge." For example, you do not need to provide a citation if you explain the difference between serif and sans serif typefaces, unless you are using a direct quotation from one of your sources. This distinction qualifies as common knowledge for those in the field of technical communication—it is something that anyone who has studied technical communication knows.

The problem is that common knowledge is not always common. For example, unless you are a wizard at English-metric conversions, you probably do not know that an 8.5 × 11 inch page size is 21.6 × 27.9 cm, but this equivalence also qualifies as common knowledge because it is a simple matter of converting from one established unit of measure to another.

Summary

Primary research involves formulating and testing a hypothesis, collecting information through observation, or conducting a survey, and then reporting the results to others in the field. Such research is called *primary* because it adds to the existing body of knowledge in the field. Secondary research is conducted by reading reports of previous research, analyzing the results, and then formulating a synthesis of the current state of knowledge on a topic. This research is called *secondary* because it draws on reports of previous primary and secondary research.

Reporting on the previous research on your topic helps you master your subject and provides your audience with the basic information they need to know to understand your research question and the context for your own research. It also demonstrates the gaps in previous research. Finally, your literature review helps establish your credibility as an expert on your research area.

An annotated bibliography provides a listing of primary and secondary sources, along with descriptive and evaluative notes about them. Although annotated bibliographies may be published in journals, books, or Web sites, they are often assembled by researchers for their own reference as a part of their background investigation for a primary research project.

A literature review is always prepared for an audience other than the researcher. It is typically a section of a research article or paper that presents a synthesis of the most important primary and secondary research relevant to the primary research project about which the article or paper is reporting. Rather than being organized by source like an annotated bibliography, the literature review is organized by topic.

References

Alred, G.J. 1997. *The St. Martin's Bibliography of Business and Technical Communication.* New York: St. Martin's Press.

American Psychological Association. 2001. *Publication Manual of the American Psychological Association*, 5th ed. Washington, D.C.: American Psychological Association.

Hofstede, G. 1994. *Culture and Organizations: Software of the Mind.* London: HarperCollins.

———. 2001. *Culture's Consequences: Comparing Values, Behaviours, Institutions, and Organizations Across Nations*, 2nd ed. Beverly Hills, CA: Sage.

University of Chicago. 2003. *The Chicago Manual of Style.* 15th ed. Chicago, IL: University of Chicago Press.

Answer Key

Exercise 3.1

APA Style

Campbell, K.S. (1990). *Theoretical and pedagogical applications of discourse analysis to professional writing.* Unpublished doctoral dissertation, Louisiana State University.

Fisher, L. (2001, January). Technical communicators: Designing the user experience. *Intercom*, 21–23.
Fishman, T. (2002). As it was in the beginning: Distance education and technology past, present, and future. *Kairos*. Retrieved March 8, 2006, from http://english.ttu.edu/kairos/7.3/binder2.html?coverweb/fishman/index.html.
Kostelnick, C., & Roberts, D.D. (1998). *Designing visual language: Strategies for professional communicators*. Needham Heights, MA: Allyn & Bacon.
Morkes, J., & Nielsen, J. Concise, scannable, and objective: How to write for the Web. Retreived March 8, 2006, from http://www.useit.com/papers/webwriting/writing.html.
Porter, E. (2004, April 28). Send jobs to India? Some find it's not always best. *New York Times*, p. A-1.
Ramey, J. (2000). Guidelines for Web data collection: Understanding and interacting with your users. *Technical Communication 47*, 397–410.
Teich, T., Redish, J.C., & Rainey, K.T. (1995). Interact to produce better technical communicators: Academia and industry. In *Proceedings of the STC 42nd Annual Conference*, pp. 57–60. Arlington, VA: Society for Technical Communication.
Usability Professionals' Association. Usability body of knowledge. Retrieved March 8, 2006, from http://www.usabilitybok.org/.

Chicago Style

Campbell, K.S. 1990. Theoretical and pedagogical applications of discourse analysis to professional writing. PhD diss., Louisiana State Univ.
Fisher, L. 2001. Technical communicators: Designing the user experience. *Intercom* (January), 21–23.
Fishman, T. 2002. As it was in the beginning: Distance education and technology past, present, and future. *Kairos*. http://english.ttu.edu/kairos/7.3/binder2.html?coverweb/fishman/index.html.
Kostelnick, C., and D.D. Roberts. 1998. *Designing Visual Language: Strategies for Professional Communicators*. Needham Heights, MA: Allyn & Bacon.
Morkes, J., and J. Nielsen. Concise, scannable, and objective: How to write for the Web. http://www.useit.com/papers/webwriting/writing.html.
Porter, E. 2004. Send jobs to India? Some find it's not always best. *New York Times*, April 28, 2004.
Ramey, J. 2000. Guidelines for Web data collection: Understanding and interacting with your users. *Technical Communication* 47, 397–410.
Teich, T., J.C. Redish, and K.T. Rainey. 1995. Interact to produce better technical communicators: Academia and industry. In *Proceedings of the STC 42nd Annual Conference*, 57–60. Arlington, VA: Society for Technical Communication.
Usability Professionals' Association. Usability body of knowledge. http://www.usabilitybok.org/.

Exercise 3.2

Because the annotated bibliography produced for this exercise will be unique for each person who prepares it, there is no key for this exercise.

4

Conducting a Quantitative Study

Introduction

If you have chosen to conduct an empirical research project, you will be collecting numerical data, and you will need to analyze that data appropriately, being careful not to draw unwarranted inferences. Or you may be reading research that uses statistics, in which case you need to think critically about whether or not the researchers were justified in making the inferences they did. The purpose of this chapter is to help you understand the principles that define "good" research when numbers are involved. Without a doubt, this is a huge subject and one that cannot be covered in a single book, let alone a chapter. We will try to keep the topic manageable in several ways.

First, this chapter limits its detailed discussion of statistical analysis to hypothesis testing involving numerical averages. Hypothesis testing is a common method in research, and most people can easily relate to concepts and calculations involving averages. Second, this chapter tries to manage the complexity of statistical calculations by not discussing the mathematical formulas involved. Instead, it discusses the principles that underlie these formulas and shows how to use Microsoft Excel to do all the required calculations.

At the end of the chapter, we discuss some other types of quantitative designs you might encounter in technical communication research articles. Those discussions are aimed at making you a better-informed reader of those articles, but they do not attempt to teach you how to conduct such analyses.

Learning Objectives

After you have read this chapter, you should be able to do the following:

- Identify examples of quantitative data
- Define standards of rigor for quantitative studies
- Differentiate between descriptive statistics and inferential statistics
- Describe the key factors that affect reliability in inferential statistics
- Write a test hypothesis and construct its associated null hypothesis for a difference in means
- Describe at a general level an analysis of variance (ANOVA), a Chi-square test, and a coefficient of correlation

Quantitative Data

Quantitative data is data that can be expressed in numbers, that is, where the observed concept of interest can be counted, timed, or otherwise measured in some way. Defining a concept so it can be measured is called *operationalizing* that concept. For example, Nielsen (1993) offers a list of items that could be quantified to operationalize the concept of usability:

- Time to complete a task
- Number of tasks that can be completed within a given time limit
- Ratio of successful interactions to errors
- Time spent recovering from errors
- Number of user errors
- Frequency and time spent using manuals and online help
- Frequency with which the manuals or online help solved the problem
- Proportion of positive to negative statements
- Number of times the user expressed frustration or joy
- Proportion of people who prefer the test system over an alternative

The following are examples of quantifiable data that could be used to describe technical writers:

- Age
- Salary
- Education level
- Years of experience in the field
- Ratios of writers to developers within companies

The following list identifies quantifiable data that could be used to describe documents:

- Document length
- Word length
- Reading grade level (RGL)
- Fog index
- Percentage of passive voice

Even somewhat abstract concepts such as reader confidence that a procedure has been done correctly can be operationalized by asking test subjects to provide a numerical rating in response to a question. For example, the usability of a manual could be operationalized by asking study participants to rate the user-friendliness of the manual on a scale of 1 (not very user-friendly) to 5 (very user-friendly).

Standards of Quantitative Rigor

In our earlier discussions, we said that research draws conclusions or inferences about a general subject or group based upon data taken from a smaller sample. Whenever inferences are made based on measurements taken from a sample, two standards of rigor must be addressed:

- The *validity* of the *measurement*
- The *reliability* of the *inference*

Validity

Validity can be divided into two types: internal validity and external validity. *Internal validity* addresses the question, "Did you measure the concept you wanted to study?" For example, consider a study that wants to compare reader preference between two different Web sites that present similar information in different ways. The researcher might decide to operationalize the concept of "reader preference" by measuring the number of times each site is visited over a fixed period of time, thus equating "more visited" with "more preferred." A critical reader of the research could argue that the test lacked internal validity because the frequency of visits could be affected by many factors other than preference—for example, differences in media promotion of the two sites or a poorly worded link to one of the sites. In other words, a measure of *frequency visited* would not be a measure of *preference.*

External validity addresses the question, "Did what you measured in the test environment reflect what would be found in the real world?" For example, a researcher might conduct a test on the effects that highly graphical, interactive design techniques have on users' satisfaction ratings of Web sites. The researcher might bring users into a lab and have them use two different versions of the same site—one rendered as plain HTML pages and the other using JavaScript, Flash, embedded videos, and other highly interactive programmatic techniques. And let us say that the study showed that users rated the highly interactive version more highly than the plainer one. But if the test was conducted on high-resolution screens, with superfast response times, and with the required plug-ins already installed, the test could lack external validity because the test platform might not represent what a typical user would have. It might be that in real life, with lower screen resolutions, longer response times, and the aggravation of loading plug-ins, the more interactive site might have resulted in much lower satisfaction ratings.

Validity is controlled through test design:

1. Internal validity can be managed by taking care when you operationalize a variable to make sure that you are measuring a true indicator of what you want to study. Ask yourself whether other factors produce or affect the measurements you intend to take but are not related to the attribute or quality you are studying. If so, your study might lack internal validity.
2. External validity can be managed by taking care when you set up the test that the conditions you create in your test environment match those in the general environment as much as possible. This design element includes ensuring that the sample group itself is a fair representation of the general population of interest. (We discuss the effect that sampling can have on test design later in the chapter.)

Exercise 4.1: Managing validity

In the following examples, discuss issues that could affect internal and external validity:

1. A company wants to compare ease of installation between one of its consumer products using a Quick Install Guide and the same product without the Quick Install Guide. It uses employees from its help desk as the subjects and times how long it takes them to install the product.
2. A study uses a sample of graduate students in a technical communication degree program to estimate how willing users will be to access a proposed online help file in an application to be used by emergency room admissions personnel. The test provides written case studies, and users are observed as they enter data into the application. The test will measure the time to complete the task and the number of errors made. ■

Reliability

Reliability describes the likelihood that the results would be the same if the study were repeated, either with a different sample or with different researchers. Obviously, a researcher wants to be able to apply the findings of the study outside the scope of just the sample observed; otherwise, there would be little value to the study. The issue of quantitative reliability is one of *statistical significance*. Statistical significance is a measure of the degree to which chance could have influenced the outcome of a test. For example, assume that one group of test subjects installs one version of a product in an average of ten minutes, and another group of test subjects installs a second version in an average of six minutes. Assuming that the same environment and setup were used for both tests, then either one of the following could be true: (1) the second version installs more quickly or (2) the researcher happened to recruit faster users for the second test. Reliable quantitative research tries to ensure that any differences detected are actually due to the object of the study and are not caused just by differences in the samples.

An Overview of Statistics

We talk about two different kinds of statistics: descriptive and inferential. *Descriptive statistics* describe a specific set of data. For example, you could record the education level of the attendees at a technical communication conference and calculate that their average education level was 16.8 years. That average would be a descriptive statistic because it describes that specific data set.

Inferential statistics, on the other hand, makes inferences about a larger population based upon sample data. For example, we might learn that the average education level for attendees at an instructional designer's conference was 16.2 years (another descriptive statistic). It would be tempting to infer from those two descriptive statistics that the average education level of technical communicators *in general* is higher than the average for instructional designers *in general*. But you cannot make such leaps unless inferential statistical techniques are applied to help you gauge the reliability of those inferences. The following illustrative example looks at how statistics helps manage the uncertainties of making inferences based on data taken from samples.

Table 4.1 Registration times (in minutes)

Participant	Version 1	Version 2
1	7.2	
2	5.6	
3	10.3	
4	6.6	
5	8.9	
6		8.3
7		11.2
8		4.6
9		6.4
10		4.3

Registration Time: Jumping to an Unreliable Inference

Bill, an information designer for a health-care provider, conducted a usability study of his company's Web site. An important feature of the site was its ability to personalize the content based upon the user's personal health profile. This personalization feature, however, depended on getting visitors to register with their personal health information on the site, so part of the test looked at problems users could have with registration. Along with a lot of qualitative data observed and analyzed, the test recorded the amount of time it took each participant to register. Five users were tested, and the product went through some revisions to eliminate problems the test had detected. Bill then tested the new design using the same scenarios and five new participants, and once again the test participants were timed. Table 4.1 shows the numerical data (time to register) taken from the two tests.

Predictably, someone in management asked, "Did the new version work any better than the first one?" Based on the qualitative results of the test, Bill was confident that the second version was easier to use; namely, users better understood the value of registering and getting their pages personalized, showed greater willingness to do the registration, and had less difficulty understanding the questions they were being asked to answer. But because "managers love numbers," Bill decided to look at the registration times from the tests to see if they also would support the claim that the second version was better—in this case meaning faster.

Just looking at the raw data in Table 4.1, it is hard to draw any immediate conclusions. Some scores for version 1 are better than some scores for version 2 and vice versa. To help understand the numbers a bit better, Bill calculated the average registration time for the "before" and "after." He found that the average installation time for version 1 was 7.72 min and the average for version 2 was 6.96 min—a respectable difference of .76 min. He felt good, then, that the quantitative data also supported that the second version had indeed done better than the first.

But when Bill presented the quantitative results to Ann, the product manager, she seemed a bit skeptical. "How do you know that you just didn't luck out and get some faster users in the second test?" she asked. She saw Bill's face go blank. She started entering the test times into her laptop computer and said, "Let me ask it another way. Let's say you tested the same version twice. What would be the odds of the second test going faster just by the luck of the draw? You know, you got some smarter users in the second group or a couple of slow-pokes in the first?"

"I don't know," Bill answered, "pretty small, I'd think."

Ann hit the Enter key on her laptop. She turned back to the Bill and looked over her reading glasses. "About 32 percent, actually," she said. "The probability is 32 percent that I could get those same results or bigger just by testing the same product twice using two different samples of users."

Bill was crestfallen.

"So now how confident are you that the second version is faster than the first?" Ann asked.

"Not very," Bill conceded. "May I ask how you came up with that number?"

Ann smiled. "Actually, I did it in a spreadsheet using one of its built-in formulas."

"It's just that by looking at the test results, the difference seemed pretty significant," Bill said.

Ann agreed, "The difference does look significant from a practical perspective, a little over 45 seconds. Who wouldn't be happy to knock 45 seconds off of any task? But the computer was looking at *statistical significance*. That means, what are the odds this difference is real rather than just being caused by sampling error?"

Bill went on, "But you seemed like you had some questions about the results even before you put the numbers in the computer. What were you looking at?"

"The same thing the computer looked at. For one thing, the sample sizes were very small. If you only test five people, the odds of an unusual user throwing off your average is a lot higher than if you test a lot of people. Second, the numbers were all over the place; in other words, there was a lot of variability in the data. When data varies a lot, the odds of picking an uncharacteristic group of fast or slow users are higher than if the data is bunched up tighter." Ann smiled again, "Of course, I couldn't come up with 32 percent off the top of my head, but that's what computers are for."

The Basic Principles

The product manager Ann in the previous example above demonstrates how statisticians think about numbers—and how you need to think about numbers when conducting or reading quantitative research. One of her questions gets to the very core of inferential statistical analysis: "Let's say you tested the same version twice. What would be the odds of the second test going faster just by the luck of the draw?" This is essentially the question of reliability and is the cornerstone underlying hypothesis testing. Inferential statistics contains sophisticated formulas and models to answer that question, formulas that are beyond the scope of this book to present and explain. But two basic principles underlie all of these formulas, two principles even a beginning researcher can and should grasp.

Principle 1: The Smaller the Variance in the Data, the More Reliable the Inference. Formulas that estimate statistical significance (that is, formulas that assess the reliability of an inference) consider how much variance there is in the sample data. Imagine for a moment that you had to estimate the weight of two different populations based on samples taken from each. Let us say that the first population is the city of Atlanta, Georgia, and the second population is all the members of the men's collegiate swimming teams in the United States. Furthermore, let us say you can take only a very small sample, ten people randomly drawn from each population. In the case of the Atlanta population, there is a lot more variance in the data than in the case of the men's collegiate swimming teams. Atlanta's population includes infants, young children, big people, small people, and so on. The swimming teams have a much smaller variance—pretty much physically fit males between the ages of 18 and 21. So the odds are much higher that you can get an "unlucky draw" in your Atlanta sample than in your swimming team sample. You could get a lot of children (making your sample average too low) or a greater than normal proportion of heavy adults (making your sample average too high). You could get unlucky draws in the swimming team sample as well, but the effects would not be as great. In other words, your lightest swimmer is not going to drag down the average that much, nor would the heaviest swimmer have much of an effect.

The point is that statistical formulas "trust" the outcome more when the data in the sample does not vary much. The more the data varies, the lower the reliability of the inference.

How do formulas take variability into account? They calculate a number called the *standard deviation*. You hear and read this term a lot in research reports, and you will probably report it in your own study. It is an indicator of the variation of the data in the sample. If the population is a typical one, you could take the average of whatever you measure (weight, salary, time to complete a task) and assume that roughly two-thirds of the population fall within plus and minus one standard deviation of that average. So, in the case of the men's swimming teams, if the average weight was 185 pounds and the standard deviation was 11 pounds, we would expect to find that two-thirds of the members of the men's swimming teams weighed between 174 and 196 pounds. Standard deviation is helpful for envisioning how widely the data vary from the average, and it is an important component of the formulas that calculate the reliability of an inference.

Principle 2: The Bigger the Sample Size, the More Reliable the Inference. Let us go back to our example of calculating the average weight for the population of Atlanta. If our sample size is ten and we get an unusually heavy person, our average will be pulled up too high. But if we increase our sample size to 500, two important things happen:

- The odds get better that our proportion of heavy people in the sample more closely matches the proportion of heavy people in the population.
- The effect of an unusual data point (called an outlier) is much smaller. For example, one unusually heavy person will have a much greater effect on an average that is calculated using 10 data points than on an average that is calculated using 500 data points.

Just as they do with the standard deviation, formulas that assess the reliability of statistical inferences take into account the size of the sample.

Although skilled researchers and advanced research designs can control sample size and variation in the data, most student research projects (and most practical research to a large degree) are stuck with what they get. It is still necessary to understand these principles, however, so that you do not overreach and make unreliable inferences from your data.

In research reports, standard deviation is represented by *SD*, and sample size is represented by *n*.

Testing for a Difference in Averages

Research in a field of practice such as technical communication is generally practical and often takes on the form of, "If I do *this*, will it make a difference?" In research, we call *this* an *intervention*. A common research pattern, and one whose roots are firmly planted in the scientific method, is to test an intervention between two groups: one group that gets the intervention (called the *test group*) and the other group that does not (called the *control group*). The intervention you test is called the *independent variable*, and the result you measure is called the *dependent variable*.

For example, you might want to test what effect using headings in a document has on how long it takes readers to find information. Using headings is the intervention you want to test (the independent variable). You test one group of participants using a document that does not have headings (the control group) and another group of participants using the same document but which uses headings (the test group). You give each group the same information recovery task, and you measure the time it takes each participant to find the information (the dependent variable).

At the end, you compare the average time it took the users of the document without headings to the average time it took the users of the document with the headings to see whether headings made a *statistically significant difference*. Once again, a difference is statistically significant if it is not likely caused by sampling error (unlucky selection of exceptionally fast or slow subjects in this case).

Hypothesis testing can also test conditions that do not involve interventions. For example, you might want to test whether there is a difference in time spent editing documents written by contract technical writers and company-employed technical writers. Although employee status is not technically an intervention, you could treat it as the independent variable, and you could treat time-to-edit as the dependent variable.

Registration Time: A Hypothesis-Testing Approach

Let us revisit our example where Bill tested the two Web site treatments and see how he could have done it more reliably as a hypothesis test. We will define and illustrate all the steps you would need to follow to conduct a reliable research project testing two groups or interventions. The approach we describe has a specific technical name: a *t-test of two independent means*, and it is a common research design. However, the principles this case illustrates apply to every quantitative study, regardless of the specific design.

Defining the Hypotheses

Bill should have started by stating his test hypothesis, that is, the assumption he was trying to test. A test hypothesis should clearly state the following:

- The independent variable (what the researcher will deliberately manipulate)
- The dependent variable (the measurable outcome the researcher will use to gauge the effect of changing the independent variable). If it is not clear how the dependent variable will be measured, state how it will be operationalized in the test hypothesis as well. For example, "... improve reader satisfaction (as measured by a satisfaction survey) ...."
- The direction (if any) of the hypothesized effect. For example, does the researcher expect the dependent variable to increase or decrease as a result of changing the independent variables?

Here is how Bill could state his test hypothesis:

There will be a statistically significant reduction in the mean* registration time between the original Web site design and a second design that resulted from applying information learned during usability testing.

But there is no way to prove that hypothesis with any certainty. Because humans are all different, and circumstances can vary, it is very unlikely that the two results would be anything but different, even if only by a little. Even if Bill tested the same Web site design twice, the odds are that the two average registration times would differ by some amount. The next logical question is, "But would the difference be big enough to say the difference was significant?" The problem is that there are no general formulas or good models to tell us what the odds are of finding a difference of "such and such size" if the two populations are different.†

So Bill solves this problem by restating the hypothesis in a form called the *null hypothesis* (so named because it claims there is no difference):

There will *not* be a statistically significant reduction in the mean registration time between the original Web site design and a second design that resulted from applying information learned during usability testing.

Why do researchers formulate and test null hypotheses? Because there **are** general formulas and models for calculating the odds of finding a difference in the means of two samples drawn from the same or equivalent populations. The logic of hypothesis testing is similar to the judicial philosophy of "presumed innocence." Juries must presume a defendant is innocent unless the prosecutor can introduce enough evidence to make them believe otherwise beyond a reasonable doubt. Similarly, researchers assume two groups or interventions are equivalent (no difference) unless the data introduce enough evidence to make them believe otherwise beyond a reasonable doubt.

The process runs like this:

1. We cannot say with any certainty whether an observed difference in registration time is caused by differences in the Web sites or just by sampling error.
2. Therefore, let us assume that there is no difference between the two Web sites; the average time to register (if we could test everybody) would be the same for both.
3. We will test the two sites (with sample users) and see how big a difference we get.
4. We will ask ourselves, "What are the odds that we would get a difference this big if the two designs really were equivalent?" (And we will use a spreadsheet to actually calculate those odds for us.)
5. If the odds are small enough ("We would have to be pretty unlucky to get a difference this big just by the luck of the draw"), we will *reject the null hypothesis* and be left with the assumption, "The difference in times must have been caused by differences in the Web site designs."

Analyzing the Data

In this section, we show you how to use a Microsoft Excel spreadsheet to calculate the descriptive statistics you will want to put in your report. Then we show you how to use

* From here on, we will stop using the word *average* and use the word *mean* instead. In statistics, *mean* means average.

† More accurately, to make that kind of calculation you would already have to know so much about the two groups that it would make the research unnecessary.

Figure 4.1 Data and labels entered into a spreadsheet.

ttest

	A	B	C	D	E
1	**Registration Times**				
2					
3		**Version 1**	**Version 2**		
4		7.2	8.3		
5		5.6	11.2		
6		10.3	4.6		
7		6.6	6.4		
8		8.9	4.3		
9	n				
10	mean				
11	SD				

Sheet1 Sheet2 S

that same spreadsheet to calculate the probability that the findings could be the result of sampling error.

First, Bill puts the data into a spreadsheet and types labels for the descriptive statistics he will report (*n*, the size of the samples; the *mean* for each version; and *SD*, the standard deviations of each sample). See Figure 4.1.

Bill then has the spreadsheet automatically calculate the three descriptive statistics for each version. Even though it is easy to look at the data and see that *n* equals 5, we will show you how to automate the calculation in case you work with large amounts of data. Bill follows these steps to calculate *n:*

1. Insert the cursor wherever you wish to display the answer (cell B9 in the example).
2. Select Insert | Function from the menu. Microsoft Excel displays the *Insert Function* dialog box. See Figure 4.2.
3. Select the category "Statistical."
4. Select the function COUNT and click OK. The *Function Arguments* dialog box appears. See Figure 4.3.
5. Select the fields that contain the data for one of the versions by clicking and dragging the mouse cursor.
6. Click OK.

Figure 4.2 Selecting the function to count the *n* of the data.

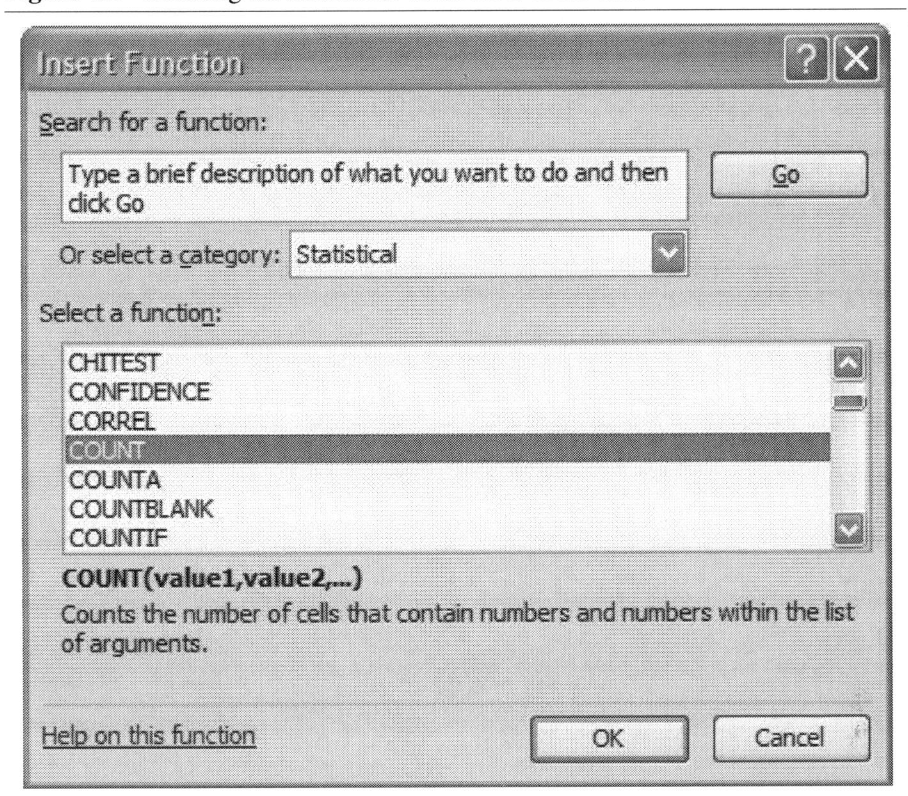

Figure 4.3 Arguments defined.

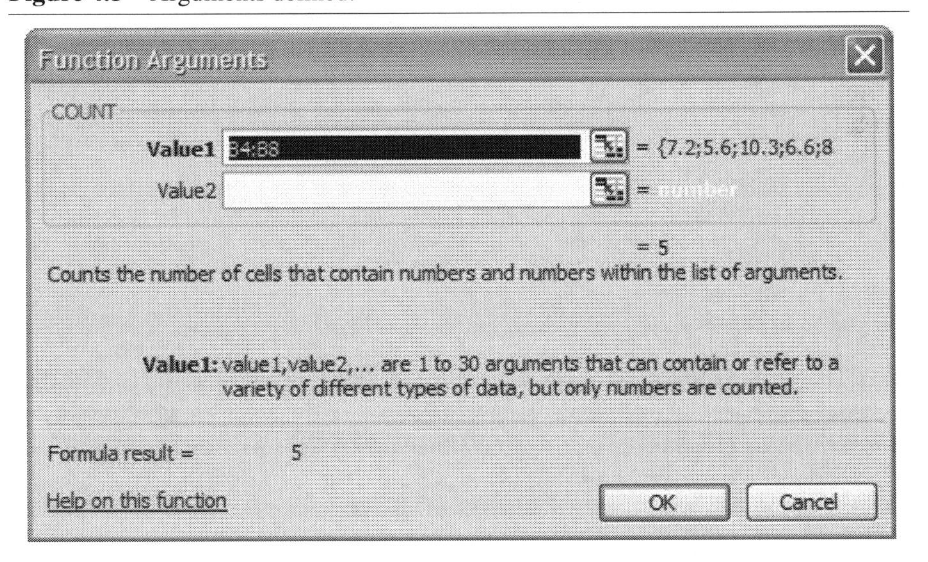

Figure 4.4 Descriptive statistics filled in.

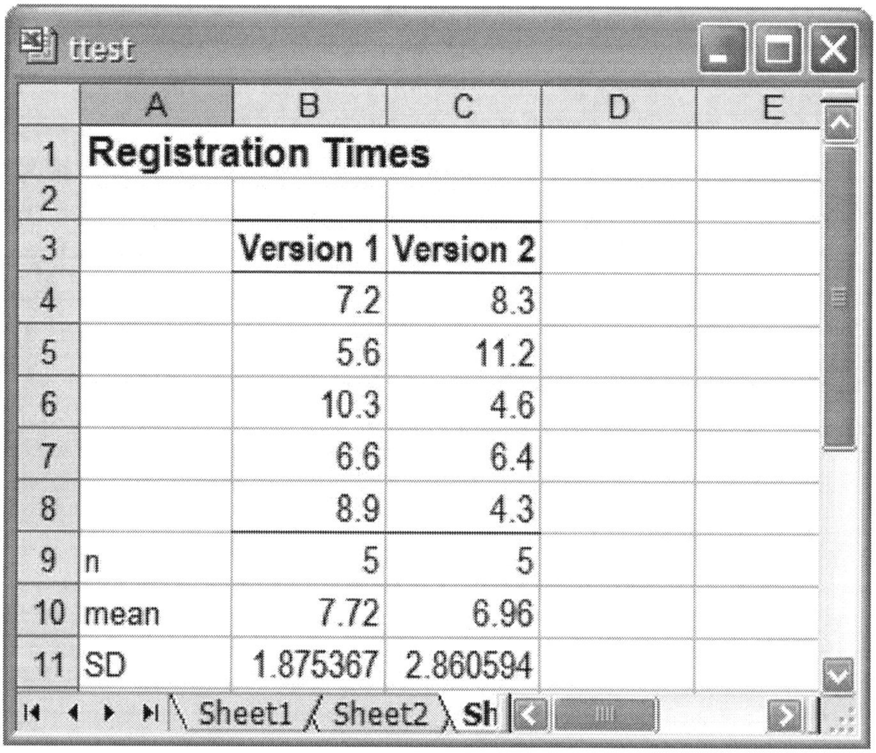

ttest

	A	B	C	D	E
1	Registration Times				
2					
3		Version 1	Version 2		
4		7.2	8.3		
5		5.6	11.2		
6		10.3	4.6		
7		6.6	6.4		
8		8.9	4.3		
9	n	5	5		
10	mean	7.72	6.96		
11	SD	1.875367	2.860594		

Sheet1 / Sheet2 \ Sh

Bill follows the same procedure to fill in the other descriptive statistics. For the mean he uses the function AVERAGE, and for the standard deviation he uses the function STDEV. (Note: Be careful to select only those cells that contain the test data. For example, do not accidentally include the cells that contain the *n* and *mean* values when calculating the *SD*.) See Figure 4.4.

Bill then uses the spreadsheet function *t*-test to calculate the probability of getting a difference in means this large from two equivalent populations. (Remember in hypothesis testing we are seeing if we can reject the null hypothesis.) See Figure 4.5.

In the *Function Arguments* dialog box Bill selects the version 1 dataset for Array 1 and the version 2 dataset for Array 2.

The tails argument (value can be either 1 or 2) refers to a one-tailed or two-tailed test. Use a one-tailed test if the test hypothesis describes an intervention that you believe will cause a difference in a particular direction (one-tailed tests are also called *directional* tests). A two-tailed (or *nondirectional*) test is one in which the test hypothesis would be supported by a difference in either direction. For example, if you were doing a study to test a hypothesis that men and women differ in regard to how much time they spend reading help files, you would use a two-tailed test because it would not make any difference in your finding if men spent more time or less time than women; a difference in either direction would support your test hypothesis that they spend a different amount of time than women do. On the other hand, if your test hypothesis was that men spend less time than women do, then you would use a one-tailed test, because the hypothesis would be supported only if the difference were in the direction of less time.

Figure 4.5 Function arguments for *t*-test.

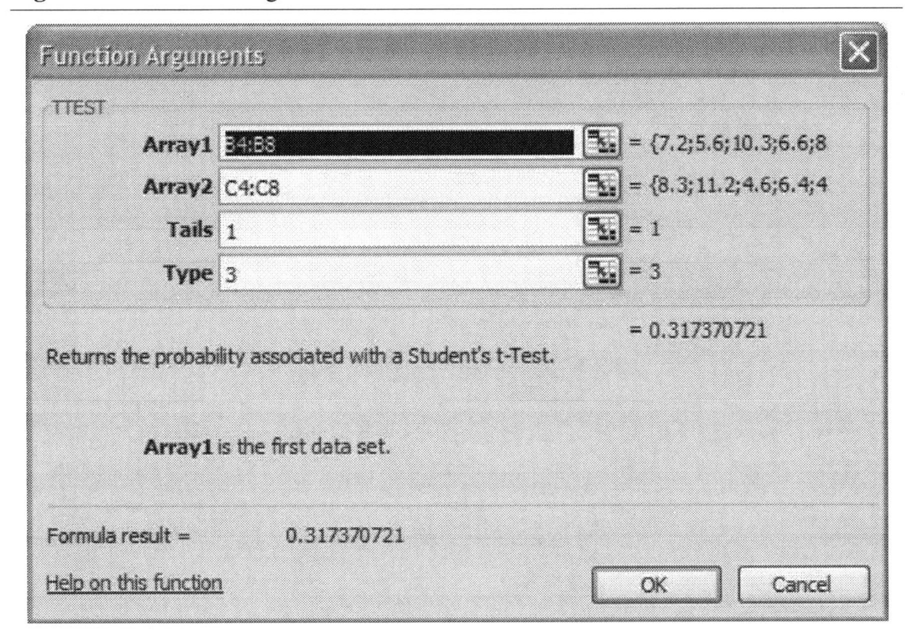

In our registration time example, Bill's test hypothesis stated that the changes in the design would reduce the mean time to register, so he selects a one-tailed test.

The Type argument defines what kind of *t*-test you are running (value can be 1, 2, or 3):

- Use "1" if the data points are paired. This type of *t*-test is common in pretest and posttest studies in which the participant takes a test, then experiences the intervention (such as instruction or training), and then takes the same test again.
- Use "2" on nonpaired tests where the variances are equal.
- Use "3" on nonpaired tests where the variances are not equal.

In the registration example, Bill makes the more conservative choice of "3." Selecting a "3" makes it slightly harder to reject the null hypothesis.

Bill labels the result of the *t*-test *p*, which stands for probability. See Figure 4.6.

Now comes the deciding question, "When is *p* small enough to reject the null hypothesis and accept the test hypothesis?" In other words, when can you say, "The odds of getting these results by chance are so small that I'm going to assume there is a real difference in the two interventions." In general, this depends on how confident you want to be. If you are not conservative enough (that is, you accept too high a *p* value), you risk accepting false results. If you are too conservative (that is, you insist on too low a *p* value), you risk rejecting true results.* Most research in the field of technical communication would be rigorous enough if it accepted results that had a *p* value of 0.1 or lower. For more conservative situations in which considerable harm could be done by accepting a false claim (for example, scrapping an expensive documentation system already in place) then a *p* value of less than 0.05 or even 0.01 might be more appropriate.

*Researchers call this type of error a Type II error.

Figure 4.6 Probability *p* calculated.

ttest

	A	B	C	D	E
1	**Registration Times**				
2					
3		**Version 1**	**Version 2**		
4		7.2	8.3		
5		5.6	11.2		
6		10.3	4.6		
7		6.6	6.4		
8		8.9	4.3		
9	n	5	5		
10	mean	7.72	6.96		
11	SD	1.875367	2.860594		
12					
13	p	0.317371			

Sheet1 / Sheet2 / Sl

Finally, when Bill writes up his final report, he copies the table, the descriptive statistics, and the *p* value into his report, along with his conclusion that there is insufficient evidence to assume that version 2 is any faster than version 1.* Actually, a lot of research ends with the conclusion that there is *no statistically significant difference* between the two groups or interventions tested. That is OK; it does not discount the value of the study. The field of technical communication still benefits from learning that two groups are not different or that a specific intervention does not have the effect that some have thought it would. Sometimes, however, the researcher thinks the observed trend would be significant if the sample size were increased and recommends that additional research be conducted with larger samples. The critical reader must weigh whether this is a reasonable suggestion on the part of the researcher or just sour grapes that the study did not produce the hoped-for results.

Checklist for Hypothesis Testing

Use this checklist to design and conduct a hypothesis test involving a difference of two means:

* Do not feel too bad for Bill; he still had a lot of good qualitative data that said version 2 was better; he just cannot say it is *faster*. We discuss how to analyze qualitative data in the next chapter.

1. State the test hypothesis. Be sure to identify the independent variable, the dependent variable, and the expected direction (if any).
2. Recast the test hypothesis as a null hypothesis. This is a formal step but it keeps everyone focused on what the test is truly looking at.
3. Collect the data.
4. Enter the data into a spreadsheet in separate columns.
5. Use the function COUNT to tally the sample sizes for each group.
6. Use the function AVERAGE to calculate the mean for each group.
7. Use the function STDEV to calculate the standard deviation for each group.
8. Use the function TTEST to calculate the *p* value, that is, the probability that the results could be caused by differences in the samples, not the intervention.
9. Decide to accept or reject the null hypothesis based on the *p* value (Typically, you can reject the null if the *p* value is less than 0.1).
10. If you can reject the null hypothesis, then accept the test hypothesis.

Exercise 4.2: Testing the difference between two means

Jane does a baseline study to determine how long it takes a help-desk person to locate part numbers for customers trying to place orders with her company. She then redesigns the parts catalog to make it more efficient and collects data again, this time using the new design. Here are the part number search times (in minutes) for the two versions:

Version 1: 1.2, 1.3, 4.5, 3.7, 2.5, 1.7, 4.2
Version 2: 2.7, 2.22, 2.52, 0.72, 0.78, 1.5, 1.02

Did Jane's redesign significantly reduce the time it took to look up part numbers?

- Write the test hypothesis you would apply.
- Write the appropriate null hypothesis.
- Calculate *n*, *mean*, and *SD* for both samples.
- Calculate the *p* value for the *t*-test.
- State what your conclusion would be if you were Jane. ■

Sampling

We have already discussed the importance of sampling for the rigor of a study.

- The external validity of the study is dependent on how realistically the profiles of the test subjects match the profile of the population of interest.
- The reliability of the study is affected by the size of the sample.

We will now discuss another important aspect of sampling called *random selection*.

In a perfect test, the only factor that would affect the dependent variable would be the manipulation of the independent variable. For example, in the case of the Web site usability test, Bill would have preferred the only factor to affect the time to register to be the respective merits of the two designs. But other factors can also introduce differences in performance:

- The innate abilities of a test subject—for example, intelligence, manual dexterity, and so forth
- The experience and education of the test subject
- Condition of the subject: for example, whether fresh or fatigued, happy or sad, and so on
- Environmental factors—for example, comfortable room conditions for some subjects against uncomfortable conditions for other subjects

The test design can try to control these *confounding variables*, as they are called, but even the best design will be subject to variances that have nothing to do with the intervention. Rigorous test methods deal with this problem by assigning test subjects randomly to the test groups. The theory is that the confounding variables will be evenly distributed across the groups and will, therefore, even themselves out.

Random selection means that the selection and assignment of each test subject is *independent* and *equal*. By *independent* we mean that selecting one subject to be in a particular group does not influence the selection of another. By *equal* we mean that every member of the population has an equal chance of being selected. In reality, it is challenging and often impossible to meet both these criteria perfectly. Researchers must often make compromises, and they need to consider (as well as disclose to the reader) how those compromises could affect the validity and reliability of the findings.

The hardest goal to achieve is to randomly select from the total population. Generally, researchers are limited to groups that they have access to. This limitation can be geographic, social, professional, and so forth. Many academic research projects, for example, rely on the student population for their sample pool. Research done by companies often relies on employees or on customer data that has been collected for another purpose. These samples are called *samples of convenience* and compromise the principle that every member of the population should have an equal chance of being selected. In those cases, the researcher should either demonstrate how the sample is still a representation of the general population or at least temper the findings with appropriate limitations.

In fact, the use of samples of convenience in an earlier study can be a good source for future research to see whether different sample groups would respond differently. For example, a study might recruit its subjects from a rural community because of the location of the university that is sponsoring the study. This test population could prompt another researcher to do a similar study with urban subjects to see whether the results would be the same.

Even if you are limited to a geographic area or by some other constraint, you can be fairly rigorous in assigning subjects to one intervention or another. Find a method that eliminates researcher or participant bias in assigning participants to groups. For example, if you are using your classmates to test the effectiveness of one type of search engine vs. another, you could use an alphabetical class roster and arbitrarily assign every other student to one type of search engine and the other half to the other type. Or you could put the names in a hat and draw them out in equal halves and make the assignments that way. Avoid letting subjects self-select or assign themselves to a group. This method could compromise the integrity of the sample.

However you select and assign your subjects, tell the reader what you have done and discuss any limitations your method might impose on the reliability of the findings.

Other Types of Studies

Based upon what you have learned in this chapter, you should be able to conduct credible research that compares the performance of two groups in terms of some variable that can be measured and averaged. You may encounter other types of quantitative research in your readings, however, and so this section introduces you to three common forms. Specifically, we describe a chi-square, an ANOVA, and a coefficient of correlation study.

Chi-Square

Chi-square analyses are often used when the variable of interest has been operationalized as a percentage. For example, a researcher might question whether the writing styles of technical communicators differ between journal articles and user manuals. One of the test hypotheses might be that there is a difference between the percentage of passive voice sentences found in journal articles and that found in user manuals. In this case, the appropriate method to test for a statistically significant difference in percentages is the chi-square test.

The chi-square test discussed in the preceding text would be called an *independent samples chi-square test*, and it is conducted in a way very similar to the *t*-test of independent means we explained in this chapter. It would test the null hypothesis: "There is no statistically significant difference between the percentage of passive voice sentences found in technical communication journal articles and that found in user manuals." Like the *t*-test of independent means, it would calculate the probability that a difference between two samples could be the result of sampling error. The main difference is that the *t*-test of independent means looks at averages and the independent samples chi-square test looks at percentages. The calculation methods are quite different, however, and it is beyond the scope of this book to go into that level of detail for chi-square tests.

Another common chi-square test is the *goodness-of-fit test*. In that kind of chi-square test, a percentage calculated from a single sample is compared to a percentage calculated from a population or some other expected value. For example, a researcher might want to know whether minorities enroll in technical communication degree programs in the same proportion that they enroll in other degree programs. The researcher could compare the percentages of minority students enrolled in a university's technical communication program and compare it to the percentage of minorities enrolled in the university as a whole. The two main differences between a goodness-of-fit test and the independent samples chi-square test are the following: (1) only one sample is taken in a goodness-of-fit test and (2) the null hypothesis in a goodness-of-fit test states the expected value. For example, if in the minorities study the university records showed that 16 percent of the students enrolled in the university were minority students, then the null hypothesis would state that 16 percent of the students enrolled in the technical communications program would be minority students (in effect, saying there is no difference between the two populations).

ANOVA

ANOVA stands for Analysis of Variance and describes a technique used when several interventions or degrees of intervention need to be tested at the same time. ANOVA studies can be a *one-way analysis* of variance or a *two-way analysis* of variance.

In a one-way analysis of variance, a single variable is manipulated as the intervention, but various manipulations are measured among different samples. For example, a researcher might want to examine the effect that type size has on reader reaction to a document. But instead of wanting to compare just two type sizes, for example, 10-point type to 12-point type for a given font, let us say the researcher would like to study the differences at 8-point, 10-point, 12-point, and 14-point types. Obviously, this design would not lend itself to the test of two means that we looked at earlier in this chapter, because four means would be involved. One could use the *t*-test of two means to test all the possible combinations of two type sizes (8 vs. 10, 8 vs. 12, 8 vs. 14, 10 vs. 12, 10, vs. 14, and 12 vs. 14) but for mathematical reasons, the reliability of so many combinations would be quite low. The ANOVA method uses a different analytical technique that gives reliable results in a test like this, with a smaller overall *n* than the *t*-test method would require.

A two-way analysis of variance allows even more leeway, allowing the tester to examine multiple variables as well as multiple manipulations. For instance, our researcher in the earlier example might want to test those four type sizes using different typefaces, such as Times New Roman, Verdana, Arial, and MS Comic Sans. A two-way ANOVA can handle such complex scenarios. It would also enable the researcher to discover interactions such as whether one type size is preferred for one typeface and another is preferred for a different typeface.

It is beyond the scope of this book to prepare you to conduct an ANOVA analysis, or even critically evaluate one that another researcher has published. However, if the research study was done as part of the writer's doctoral thesis, (a common source of ANOVA-based research articles), you can feel confident that it underwent a rigorous review by the writer's doctoral committee.

Coefficient of Correlation

Some studies do not make interventions, strictly speaking, but just look to see whether a relationship exists between two variables. For example, a researcher might want to see whether there is a correlation between time to install a product and the perceived ease of installation. This would not necessarily involve an independent variable and dependent variable situation. The researcher could just have a number of test subjects install the same product, time them, and then have them rate the ease of installation. The researcher could then calculate the *coefficient of correlation* to see how strong the relationship was between the time test subjects took to install the product and the subjects' rating of the ease of installation. The result of that calculation is called *r*, and its value is always equal to or between +1.0 and –1.0. If the number is positive, the relationship is said to be *direct*; that is, an increase in one variable is associated with an increase in the other. If the number is negative, the relationship is said to be *inverse*; that is, an increase in one variable is associated with a decrease in the other.

The following general guide can be used to interpret correlation coefficients:

- If *r* is less than 0.3, the relationship is a weak one.
- If *r* is between 0.3 and 0.7, the relationship is a moderate one.
- If *r* is greater than 0.7, the relationship is a strong one.

It is important to note that *r* is a descriptive statistic (and easily calculated using the function CORREL in Microsoft Excel); it describes the relationship found in the sample dataset. Just as in a test of means, the researcher needs to state whether or not the finding was statistically significant. It is also important to note that correlation studies do not show causation. They only show the degree to which one variable can be used to predict another, and not that one causes the other.

Summary

Whenever inferences are made based on measurements taken from a sample, two standards of rigor must be addressed:

- The *validity* of the *measurement*
- The *reliability* of the *inference*

Internal validity addresses the question, "Did you measure the concept you wanted to study?"

External validity addresses the question, "Did what you measured in the test environment reflect what would be found in the real world?"

Descriptive statistics describe a specific set of data—usually the sample.

Inferential statistics make inferences about a larger population based upon sample data.

The two underlying principles of inferential statistics are

- The smaller the variance in the data, the more reliable the inference.
- The bigger the sample size, the more reliable the inference.

Hypothesis testing is a way to see whether an intervention makes a difference (the test hypothesis). It tests the assumption that two groups (control group and test group) are alike. If it can reject that assumption with confidence, it accepts the alternate test hypothesis that the intervention made a difference.

Random selection means that the selection and assignment of each test subject is independent and equal.

- *Independent* means that selecting one subject to be in a particular group does not influence the selection of another.
- *Equal* means every member of the population has an equal chance of being selected.

References

Nielsen, J. 1993. *Usability Engineering*. Boston, MA: Academic Press.

Answer Key

Exercise 4.1: Managing validity

1. As far as internal validity is concerned, speed of installation could be said to be a reasonable way to operationalize at least an aspect of usability. Depending

on how broadly the study wishes to describe usability, one could argue that it should also look at other aspects as well, such as user satisfaction, confidence that the product was installed correctly, and so on. However, the selection of the product's help-desk employees as the test subjects seriously compromises the external validity of the test. These subjects bring skills and product knowledge to the experience that would not be available in the general population.

2. A serious internal validity issue in this case is that the study purports to examine "willingness to access help files" but it measures variables completely unrelated to this (time to complete task and number of errors). Its external validity is seriously flawed by the use of technical communication students, who might well have an entirely different attitude about using help files than hospital admissions personnel. Also, the external factors associated with admitting someone into a hospital (such as time pressures because of life-threatening conditions) would be difficult to reproduce in a lab environment. Not accounting for such real-world factors could compromise the external validity of the study.

Exercise 4.2: Testing the difference between two means

H_1: There is a statistically significant reduction in the time to research part numbers between the original parts catalog and the redesigned catalog.

H_0: There is *not* a statistically significant reduction in the time to research part numbers between the original parts catalog and the redesigned catalog.

Data Analysis:

	Ver 1	**Ver 2**
	1.2	2.7
	1.3	2.22
	4.5	2.52
	3.7	0.72
	2.5	0.78
	1.7	1.5
	4.2	1.02
n	7	7
Mean	2.728571	1.637143
SD	1.398469	0.839081
p	0.051001	

Conclusion: The new design had a practical impact (reduced search time to 60% of the original), and the findings were statistically significant ($p < 0.1$). Based on these results, I (Jane) propose that we use the new format for future parts catalogs.

5

Conducting a Qualitative Study

Introduction

Technical communication is essentially a social study, one that tries to understand how people communicate with each other within technical domains. Additionally, technical communicators are members of professional communities, communities of practice, and the community at large. So it is natural for research within the field of technical communication to draw heavily on qualitative methods from fields such as sociology and anthropology. There are rich and fertile opportunities for research that deals with such questions as "How do people use the communication products we create?" "How do we collaborate with others in making these products?" and "How do readers make sense of what we write?" Furthermore, there are broad social implications to what we do. For example, do we make technology accessible to those who would otherwise be disenfranchised, or do we systematically exclude classes of potential users by explaining technology using media and semantics that favor the "ins" and exclude the "outs"?

In trying to find meaningful answers to questions such as these, we find that the quantitative methods discussed in the previous chapter seem inadequate. For example, although those methods can tell us with great precision and confidence what percentage of users will not use a help file when in trouble, their precision dulls and our confidence in their accuracy falters when we ask the follow-up question, "Why not?" It is for questions like these that we turn to the disciplines and practices of qualitative research.

Learning Objectives

After you have read this chapter, you should be able to do the following:

- Identify examples of qualitative data
- Define the standards of rigor for qualitative studies:
    - Credibility
    - Transferability
    - Dependability
- Describe ways of ensuring rigor in qualitative studies
- Apply coding and categorization schemes to analyze qualitative data

Qualitative Research

Qualitative research can be defined by the type of data it relies on and by the methods it applies in gathering and analyzing that data. Qualitative data is nonnumeric—it is words, images, objects, or sounds that convey or contain meaning. Qualitative methods generally involve the following three phases:

1. The observation of behaviors or artifacts in natural surroundings or authentic contexts
2. The recording or noting of those observations in descriptive or narrative formats
3. The systematic analysis of those observations or notes to derive patterns, models, or principles that can be applied beyond the specific events or artifacts studied

The following are common qualitative methods within technical communication research:

- Interviews—what respondents say when asked (one-to-one)
- Focus groups—what respondents say when asked (one-to-many)
- Usability tests—what users do when observed in a controlled environment
- Field observations—what users do when observed in their natural environment
- Document analyses—what documents tell us about the authors or the intended audience

Standards of Rigor in Qualitative Research

Although qualitative research has different methods for ensuring rigor than does quantitative research, it is no less interested in achieving that same level of rigor. Merriam (1998) states this common concern: "All research is concerned with producing valid and reliable knowledge in an ethical manner" (198). She again reinforces the similar goals of the two methods: "Assessing the validity and reliability of a qualitative study involves examining its component parts, as you might in other types of research" (199).

But Corbin and Strauss (1990) state that many readers of qualitative reports tend to read them with a quantitative interpretation: "Qualitative studies (and research proposals) are often judged by quantitatively oriented readers; by many, though not all, the judgment is made in terms of quantitative canons" (4).

Recognizing this problem, Lincoln and Guba (1985) held that qualitative research needs a vocabulary that differentiates it from quantitative methods, and they suggested a set of terms shown in Table 5.1 to correlate qualitative concerns with their similar concerns in quantitative analysis.

Table 5.1 Comparison of quantitative terms to qualitative

Quantitative	Qualitative
Internal validity	Credibility
External validity	Transferability
Reliability	Dependability

Credibility

In a quantitative study we assess the *internal validity* by essentially asking "Did you measure the concept you wanted to study?" In a qualitative study, where data is not measurable, we look more to the *credibility* of the data; that is, do the participants truly represent the population or phenomenon of interest and how typical are their behavior and comments? For example, a researcher might study how help-desk personnel use a product's technical documentation, and for this study the researcher interviews help-desk supervisors across several companies. A critical reader of the research could challenge the credibility of the study on the grounds that supervisors are not necessarily appropriate spokespersons for how their employees actually use the documentation. Similarly, a gender-bias study that interviewed only male managers to determine whether female technical communicators were treated equitably would lack credibility.

Credibility can also be affected by the methods employed in the study, especially if the methods affect how freely and honestly respondents can participate. For example, a focus group that contains managers and line employees could compromise credibility if subordinates felt pressured to answer questions in a particular way by the presence of their managers in the group.

And finally, *observed behavior* (watching what people do) has higher credibility than *self-reports* (having them tell you what they do). It is very common in usability tests, for example, to see a conflict between user self-reports during pre- or posttest interviews and their observed behavior during the test. For example, during pretest interviews respondents might say that they rely heavily on product documentation, yet never use the documentation during the test, even in the face of difficulties in accomplishing the task. Seely Brown (1998) cites a study done at Xerox where accounting clerks were interviewed about how they did their jobs. Their individual descriptions were consistent with the written procedures; however, subsequent on-the-job observations showed that their descriptions were very different from how they actually performed the tasks. The upshot of all this is that if your research relies heavily on self-reported data such as interviews or focus groups *as a way of describing behavior*, you must be concerned about the credibility of your findings. A good guideline is to use interviews and focus groups to discover people's opinions, motives, and reactions—not to learn about their behavior. If you want to know what people *do*, you are better off watching them do it rather than asking them what they would do. But if you want to know *why they do it* or *how they feel about it*, then interviews and focus groups can be credible methods.

Transferability

In quantitative research, *external validity* addresses the question "Does the phenomenon you're measuring in the test environment reflect what would be found in the real world?" In qualitative research the question is essentially the same, with the substitution of "observing" for "measuring." The question could also be stated, "How natural or authentic was the environment in which the study took place?" Sometimes, "natural" is difficult to achieve, in which case the emphasis needs to be on "authentic." Authentic means that the context of the behavior being studied is consistent with the context of the real-world situation in which the actual behavior would occur.

For example, in the early days of usability testing, researchers worried a lot about making the lab environment feel like the respondent's natural environment. Emphasis has now shifted more to ensuring that the tasks are authentic, that is, that the users are asked to achieve realistic goals with the product. More recent usability research has started to use "interview-based tasks" where the test starts with an interview to determine what that participant might actually do with the product or Web site being tested. Then the test consists of observing the user doing that task, not one made up by the researcher. This approach could be said to have higher transferability than one that used predetermined or scripted tasks.

Researchers must often balance the ability to observe and record data in a controlled environment vs. the increased transferability of using the respondent's natural environment. The more the researcher must compromise the naturalness of the environment, the more he or she must attend to the authenticity of the tasks.

In a real sense, much of the burden for transferability falls on the reader of the research report. "Qualitative research passes the responsibility for application from the researcher to the reader" (Firestone 1993, 22). The reader must decide to what degree "these respondents and these circumstances" resemble the population and environment of interest to the reader. For example, a manager of Web-content developers might be interested in problems related to managing creative tasks. That manager might decide that a study of how graphic-arts supervisors managed creativity was very relevant to the problems she was trying to solve. On the other hand, a study of how supervisors managed the productivity of proposal writers, although seemingly more similar to technical communication, might have less relevance to her situation of managing creativity. Researchers help the readers make these decisions about transferability by describing the respondents and their circumstances with a greater emphasis on richer descriptions of the participants and their contexts than might be seen in a quantitative study.

Dependability

As with its quantitative counterpart, reliability, *dependability* refers to the confidence with which the conclusions reached in a research project could be replicated by different researchers. In qualitative studies, more so than with quantitative studies, it is more difficult to turn off the researcher's own subjectivity; therefore, qualitative researchers must be careful to ensure that their conclusions have emerged from the data and not from the researcher's own preconceptions or biases. Unlike quantitative studies, which can rely on procedural, statistical methods to do this, qualitative studies must rely on more humanistic protocols. The literature on qualitative research discusses many techniques that can be applied to verifying the conclusions produced by such research. In this discussion, we try to summarize them into the following categories:

- Depth of engagement
- Diversity of perspectives and methods
- Staying grounded in the data

Depth of Engagement

Generally, the more opportunities researchers give themselves to be exposed to the environment and to observe the data, the more dependable the findings will be. For example, a case study that looks at writers in a department over a one-day period is not going to be as dependable as one that observes the same department over a period of three months. Longer studies gather more data within a broader array of contexts and are less susceptible to what is called the Hawthorne effect, the phenomenon where

Figure 5.1 Graph of cumulative findings showing data saturation.

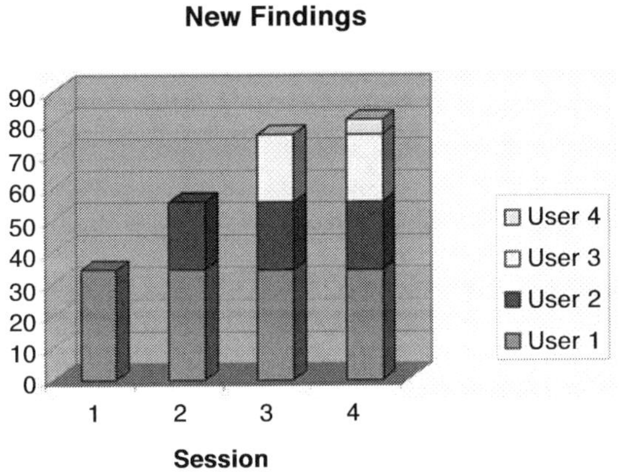

behavior might change temporarily just because of the novelty of the new treatment or the attention of the researcher.

However, depth of engagement is not necessarily measured in numbers, such as length of time or number of participants. Another gauge is the richness of the context and the study's success in achieving *data saturation*. Data saturation is the point at which staying another day, interviewing another respondent, or testing another user ceases to add any new data. Investigations of usability studies reveal that we can reliably predict that this will happen after testing four to seven users, but for many other kinds of studies, you would not know where data saturation occurs until you hit it, in essence, when you start observing the same themes or behaviors occurring with no new insight or data being added. Figure 5.1 shows a graph from a research project that included data from a usability study (Hughes 2000). The author used the graph to support his assertion that data saturation had occurred in the usability test by noting the sharp drop-off in new data after the third session. (He supported this conclusion with literature references that reported similar findings in other studies.)

Diversity of Perspectives

Qualitative research often relies on *triangulation* to demonstrate its dependability. The term comes from a technique for using a map and a compass to determine location. Essentially, it relies on comparing compass readings of a point taken from two other points and plotting their intersection on a map (you are here!). The term is also used in qualitative research to describe methods that look at data from different perspectives. Seeing data from multiple perspectives—for example, using multiple researchers or multiple data collection techniques—increases rigor.

For example, let us say a researcher does a study that relies on field observations in the workplace, one-on-one interviews with the writers, and analyses of the documentation produced, and then the researcher draws conclusions based on those three sources of data. That study would be more dependable than one that used just one of those data collection methods. Similarly, a usability study in which multiple observers watch the user will be more dependable than one that has just a single observer.

Diversity of perspectives can also be applied to the diversity of the respondents the researcher involves in the study. Here, qualitative research can differ markedly from quantitative research. In quantitative studies, the researcher often tries to reduce diversity among subjects as a way to minimize variance. (Remember from our chapter on quantitative methods that reducing variance can increase reliability.) For example, if a researcher wanted to test which of two user interfaces was quicker to use, he or she might recruit users with similar computer expertise and background. This strategy would reduce variance in the data not related to the actual treatments being tested and would make the results more reliable. But a qualitative researcher wishing to understand how users would search for information on a company intranet might deliberately recruit participants of varying experience and expertise to get a richer perspective of how the intranet would be used in general.

Similar to triangulation is the technique of *peer review*, in which multiple perspectives are gathered after the fact. In peer review, the researcher allows others to examine the data and the findings to determine whether they interpret them the same way as the researcher does. One of the vulnerabilities of qualitative research is that it can be subjective. Kidder speaks of researchers experiencing a "click of recognition" or a "Yes, of course" reaction (1982, 56) during qualitative research. This is a great strength of qualitative research, the clarification that comes from seeing something through the participant's perspective, but you must guard against the subjectivity that might have led to that "of course" reaction—one person's "Of course!" could be another person's "What the heck was that?"

Peer review helps eliminate some of that subjectivity by viewing the data through multiple subjective lenses. Consensus strengthens the dependability of the conclusion. When consensus cannot be reached, the researcher may choose to say, "Since I'm the only one who interprets this respondent statement this way, I'll drop it," or the researcher might choose to note in the report that peer reviewers saw it differently, giving the reader both perspectives.

Another way to bring validating perspectives into a research project is to use *member checking*. Member checking is a technique by which the researcher solicits the respondents' views of the credibility of the findings and interpretations. This can be done at the end of the study by asking respondents to read the report and verify the observations and conclusions the researcher has made, or it can be done within the study by the researcher validating his or her perceptions at the time with the respondents. This real-time member checking is common in interviews, focus groups, and usability tests when the researcher restates his or her understanding of what the respondents have said or done for their immediate verification or correction.

When writing up a qualitative research report, it is a good practice to describe any of the methods used to ensure the dependability of the conclusions.

Staying Grounded in the Data

Because qualitative studies do not have the controlled structure of quantitative studies, they are more susceptible to researcher bias or subjectivity. For that reason, it is very important that all conclusions and statements be traceable back to directly observed data within the study. Qualitative reports are heavily laced with quotations from respondents, samples of artifacts, or media clips of respondent behavior. Furthermore, for a qualitative study to have rigor, it must employ a formal, systematic technique for examining the data and finding patterns or common themes across the data.

The positive aspect of qualitative studies' looser structure is that they can "go where the data takes them" in ways that quantitative studies cannot. For example, in a technique known as *negative case analysis*, the researcher refines and redefines working hypotheses in the light of disconfirming or negative evidence. Here, we see the contrast between quantitative and qualitative methods. In empirical studies, variance is controlled by keeping the conditions of the test as constant as possible throughout the test. However, in qualitative studies, it is not at all unusual to change tasks, questions, and even participant profiles as you go.

This pattern of evaluating the data during the study and then modifying the course of the study is very common in *grounded theory* studies, in which models are built, evaluated, and modified after every interview, and new interview questions and respondent profiles are subsequently created to validate or test the emerging model or theory. In this kind of approach, the emphasis is not on proving or disproving a hypothesis; rather, it is on refining it at each iteration. Because the final hypothesis or model reflects a broader diversity of respondent input, it is considered more dependable than one tested inflexibly over several respondents.

In a similar application of this principle of "go where the data takes you," it is a legitimate practice in usability studies to change scenarios or even modify the prototype being tested to validate a finding or just to move on and collect more data if a particular problem has become obvious. For example, if the first two users in a study fail to use a critical link, and the observers all have the same "click of recognition" that the link has been poorly signaled, it is acceptable to reword the link and test that alternative wording with the subsequent users. Such a change would be an unforgivable breach of protocol in a quantitative study.

Because all of this can be quite subjective at worst and messy at best, it is important that qualitative studies incorporate systematic methods for examining and analyzing the large quantities of data they gather. Later in this chapter, we describe some useful qualitative data analysis techniques.

Qualitative Methods

In this section, we discuss guidelines for planning and conducting qualitative studies that use any of the following methods:

- Interview
- Focus groups
- Usability testing
- Field observation
- Document analysis

Interview

An interview is a structured conversation between a researcher and a respondent. The degree of structure is determined by the *protocol* or list of questions the researcher plans ahead of time. A good protocol usually involves a combination of open-ended questions—questions that call for long, descriptive answers from the participant, and close-ended questions—questions that call for short, validating answers.

You plan for an interview by defining the profile of the respondents you wish to interview, planning how you will recruit them, and writing out the protocol you will use. The questions you ask should be derived from the research goal and questions you have defined for your study. Some initial questions might be added as "ice breakers," that is, not so much intended to gather data related to the study as to get the respondent talking and feeling comfortable.

When conducting the interview, try to maintain the tone of a conversation. Be familiar with your questions so that you do not have to read them, and try to provide natural transitions between questions, such as "That was an interesting point you made about how the subject matter expert's time is hard to secure. I'd like to follow up on that by asking ...."

Also, do not be afraid to deviate from the protocol if the respondent makes an interesting but unanticipated comment. In that case, ask him or her to elaborate or pursue the point with a question not necessarily in your protocol. Do this, however, only if the new direction is consistent with your research goal.

A good way to close an interview is to ask the respondent if he or she has anything else to add on the topic you have been discussing that has not already come out in your conversation. This question can often unveil some rich data or insight you had not anticipated.

Focus Groups

Focus groups are essentially group interviews and should address the same issues and follow the same guidelines stated for interviews. In the case of focus groups, however, there are some additional considerations.

Make an effort to draw out diverse perspectives in nonthreatening ways. The purpose of a focus group is to get multiple perspectives, and this can sometimes be lost as some people are reluctant to offer opinions that are different from what other, more vocal participants might have offered. The best way to get at these differing opinions is to avoid positioning them as disagreements. Do not ask, "Does anyone disagree with what Mary just said?" Instead, ask, "I'd like to hear other opinions on that. Can someone share a different perspective from what Mary just described for us?"

Also, some respondents will just be less talkative than others. Be prepared to call on them directly to get their input.

Usability Testing

Usability testing is a common form of technical communication research. In usability testing, we observe users doing authentic tasks with a product or set of documentation. To plan a usability test research project, you must address the following questions:

- What is the profile of the respondents you wish to observe?
- What tasks will you have them do?

User Profile

When you are defining your user profile, you should address two main areas: technical expertise (if technical devices such as computers are involved in the tasks) and subject domain knowledge. For example, if you were studying the effectiveness of computer-based training for a nursing course, the participants would need to be nurses or nursing students, and they would have to have the same level of computer skills that the proposed student population will be expected to have.

To determine a potential participant's computer expertise, it is more reliable to ask behavioral questions than just have the recruit rate their skills as novice, intermediate, or expert. A useful technique is to identify commercially available software products that require comparable skills to the one you will be testing and ask the recruits whether they have used those products and what they have done with them. For example, if you are conducting a study that involves doing an Internet search, you could ask potential participants to identify which search engines they have used and what they searched for.

A practical note: The more detailed and restrictive the user profile, the more challenging the recruiting will be. The balance is to be detailed enough to ensure credibility and transferability without overly constraining yourself.

Tasks

The tasks should relate easily back to the research goal and questions of your study. For your study to be credible, ensure that your tasks are authentic. For example, if you are studying how users scan a results list from a search query, it would be better to state a search goal and ask the user to identify the result they found to be most relevant, such as "Find an article that discusses the differences between quantitative usability tests and qualitative usability tests" instead of asking them to "Search on 'Usability' and find the article written by Hughes."

In short, do not tell the participants how to do a task; instead, tell them what to accomplish and then observe how they go about it.

Field Observation

Field observation can be a very credible technique for qualitative research because it relies on observations made in the participants' natural environment. Of course, you have to be aware that the presence of a researcher can alter the environment and thereby detract from the credibility. Essentially, in a field observation, the researcher observes the participants going through their natural routines.

The observations can be very open, for example, a shadow study in which the researcher follows the participants and notes what they do within their normal routine. Let us say that a researcher is studying how help-desk personnel use the product's technical manuals in solving problems. One method might be to spend a day or half-day sitting with a customer-care representative as the representative takes customer calls and noting how he or she researches the answers to the problems.

The observations can also be very focused. For example, another researcher might be interested in how nonaccounting managers use a spreadsheet application to put together their department budgets. In this case, the researcher wants to be present to watch only that particular task. In both cases, open and focused, the emphasis is on observing how the participants actually go about a task or collection of tasks in their real-world setting.

Note-taking in a field observation study can be as detailed as videotape transcripts of the entire time of interest or as loose as observation notes recorded manually in the researcher's notebook. In the latter case, the emphasis should be on noting what the participant does, the sequence in which steps are performed, the length of time he or she spends on a task (it is a good idea to keep field observation notes in the form of time-based journal entries), the tools or resources he or she uses, whom he or she talks to or seeks help from, etc.

Perhaps more so than any other form of data gathering, field observations can seem the most intrusive on a respondent's or sponsor's time or sense of confidentiality. Care must be taken to gather the proper permissions—telling both the respondent and the sponsor how the data will be used and what their rights as participants are.

Document Analysis

A lot of what technical communicators do is centered on the production of various kinds of documents, and much can be learned from examining the documents that they produce. In a quantitative study, standard methods of analysis might include document length, reading grade level, percentage of active vs. passive voice, and so forth. In a qualitative study, however, you would be looking at nonquantifiable areas of interest, such as tone, style, and vocabulary selection.

For example, a quantitative study might note that technical white papers used more passive voice constructions than did user guides on the same topic (a simple analysis of frequency of passive constructions), whereas a qualitative study might note that user guides tended to shift into passive voice when potential negative consequences were described—perhaps in an attempt to distance the manufacturer from accountability or association with the negative consequences. Both are observations about how passive voice is used, but the qualitative is more interpretative.

Data analysis

Before you can analyze qualitative data, it must exist in some physical form. This form can be transcripts of the interviews you conducted, detailed logs you kept, video transcripts, or documents you are studying. Having these artifacts in electronic format allows you to manipulate and mark them up during the analysis phase. Many professional data analysis tools as well as conventional spreadsheets and word processor applications can make your analysis easier if the data is in electronic format.

Qualitative data analysis consists of the following three phases:

1. Coding
2. Categorizing
3. Modeling

Coding

Coding involves breaking the data into the smallest chunks of interest and describing it at the chunk level. How small a chunk should be depends on what you consider to be your basic *unit of analysis*. Generally, we analyze discourse (text and transcripts of respondent comments) at the sentence or phrase level. In some cases, however, it might make more sense to use larger units, such as document sections, or smaller units, such as individual words, as the units of analysis. Field observation notes are typically analyzed at a higher level, because researcher notes are not as detailed as a respondent interview transcript would be.

The codes that are applied during the coding phase can be *predefined codes* or *open codes*. Predefined codes are determined before the analysis begins and are usually taken from an existing model or theoretical structure on which the researcher is basing the analysis. For example, a researcher might be studying to what degree certain documents display the information types identified by the Information Mapping® taxonomy for structured writing. In this case, the researcher would use the information types as the codes and would code sections of the texts to reflect which type they represented—for example, *structure*, *concept*, *fact*, and so on. As another example, professional usability labs often have a standard set of codes based on established usability heuristics that they reapply from test to test, such as "Navigation" or "Terminology."

It is more common to see open codes used in qualitative analysis, especially if the purpose of the study is to identify new patterns, taxonomies, or models. In an open coding approach, the researcher creates the codes based upon the data. Many times, the open code is taken from the words in the transcript itself, a technique known as *in vivo* coding. For example, a researcher might code the respondent phrase, "I get so frustrated when the instructions assume I know where I'm supposed to be in the application," with the word *frustrated.*

When coding data, it is sometimes useful just to scan the transcript first, highlighting statements of interest, and then to go back and apply codes to comments or incidents that seem significant. This approach is especially useful for the first transcript or document you are analyzing, before you have a sense of what themes might emerge.

Perhaps the greatest challenge for new researchers is deciding how to code, that is, what should the codes look like, how long or short should they be, how much they should code. Welcome to qualitative research!

The good news is that coding is flexible and fluid. By that we mean that it can be done correctly in a number of ways and that you can always change your coding scheme as patterns or models start to emerge. In fact, this is an important strategy in qualitative analysis, so if you find yourself changing codes, this just means that your insight is emerging from the data and not from your preconceptions.

Not all statements need to be coded, only those that seem to have importance to the study, and some data can have multiple codes. In the previous example, the sentence "I get so frustrated when the instructions assume I know where I'm supposed to be in the application" could have been coded *frustration* and *location in the application* and even *instructions* if the researcher wanted to. The important thing is to dive into the data and start coding rather than stay outside the data and make broad assertions.

Electronic transcripts can be coded using special software applications designed for qualitative research, or you can use the indexing feature in a word processor. We provide an example at the end of the chapter that shows how to use a word processor as a qualitative analysis tool.

Categorizing

The next phase in the analysis of the data is to start to look for patterns or groupings in the codes. In this phase, codes are grouped into categories, and even categories can be regrouped under higher-level categories. Whereas the coding was an analysis of the directly observed data, categorizing is a more abstract analysis of the codes themselves. This is an important step in making the findings of the research transferable to other events or situations than just the one being observed.

During this phase, the researcher often rewords codes or even breaks them down into lower levels of coding, making the original code a category. For example, a researcher might have been coding certain types of respondent comments as *emotional reaction*, but then decides in the categorizing phase to break that down into *frustration, anger, relief, thrill,* and so forth. Doing this means going back into the data, finding all the occurrences of the code *emotional reaction*, and recoding them. Although this might seem to be inefficient, this bouncing back and forth is actually part of the rigor of good qualitative analysis. Once again, it is an indication that the conclusions are emerging from the data and not from the researcher's preconceptions. A helpful tip, however, is to avoid starting with abstract codes, for example, starting with the code "emotion" rather than "frustration." Your initial codes should stay close to the data, and your abstractions should emerge primarily from the categorization phase of your analysis.

Categorizing can be done manually with index cards, or you can use software products to move and manipulate the codes you have created. Graphing tools can be useful during the categorizing stage. If you use Microsoft Visio, the Brainstorming stencil allows you to enter codes, create categories, and then manipulate them graphically. Figure 5.2 shows an example of codes grouped into categories in Microsoft Visio. Note that you can see the categorization by looking at the graphic representation or by looking at the Outline Window.

For the categorizing phase to be productive, it should be iterative and experimental. Keep moving things around, renaming them, making more codes, and getting rid of codes that do not seem to work. In short, keep working with the data until it starts to fall into place. Remember that in qualitative research, the aim is *not to eliminate* the subjective insight of the researcher *but to manage* it so that it facilitates transferable, dependable conclusions.

Modeling

The modeling phase is the natural outcome of the categorizing phase, and often the one blends seamlessly into the other. The example in Figure 5.2 in which a graphical mapping tool was used to create categories is a good example. What emerged from the categorizing phase was an affinity map of what information respondents felt they needed when they paid their bills—a useful model for the researchers who were designing the screen layouts and flows for a Web-based bill-pay application.

In other cases, the emergence of patterns or models might not flow as naturally. What the researcher is looking for in this phase are those generalizations that can be applied beyond the boundaries of the particular study. These could take many forms:

- List of principles or axioms
- Table that summarize roles or relationships
- Set of design heuristics
- Process diagram
- Affinity or mind map

The exact form a model should take is up to the researcher. The criterion should be what form best communicates the type of findings. For example, Figure 5.2 shows a model of the information that respondents thought was important when they paid their bills expressed as an affinity map (mind map) to show the complex information relationships.

Figure 5.3 shows a model of the various roles a facilitator plays during a usability test shown as a table to better contrast the roles; Figure 5.4 shows a model of how team learning occurs during a collaborative usability test shown as a process diagram to illustrate inputs and outputs of that process and the events in between.

The rigor of qualitative research is due to the fact that whatever output your research creates, it can be traced back to the original data; that is, the model is derived from the categories, the categories were derived from the codes, and each code is tied to one or more instances in the data. In fact, in writing up the research report, it is common to

Figure 5.2 Categorizing in Microsoft Visio.

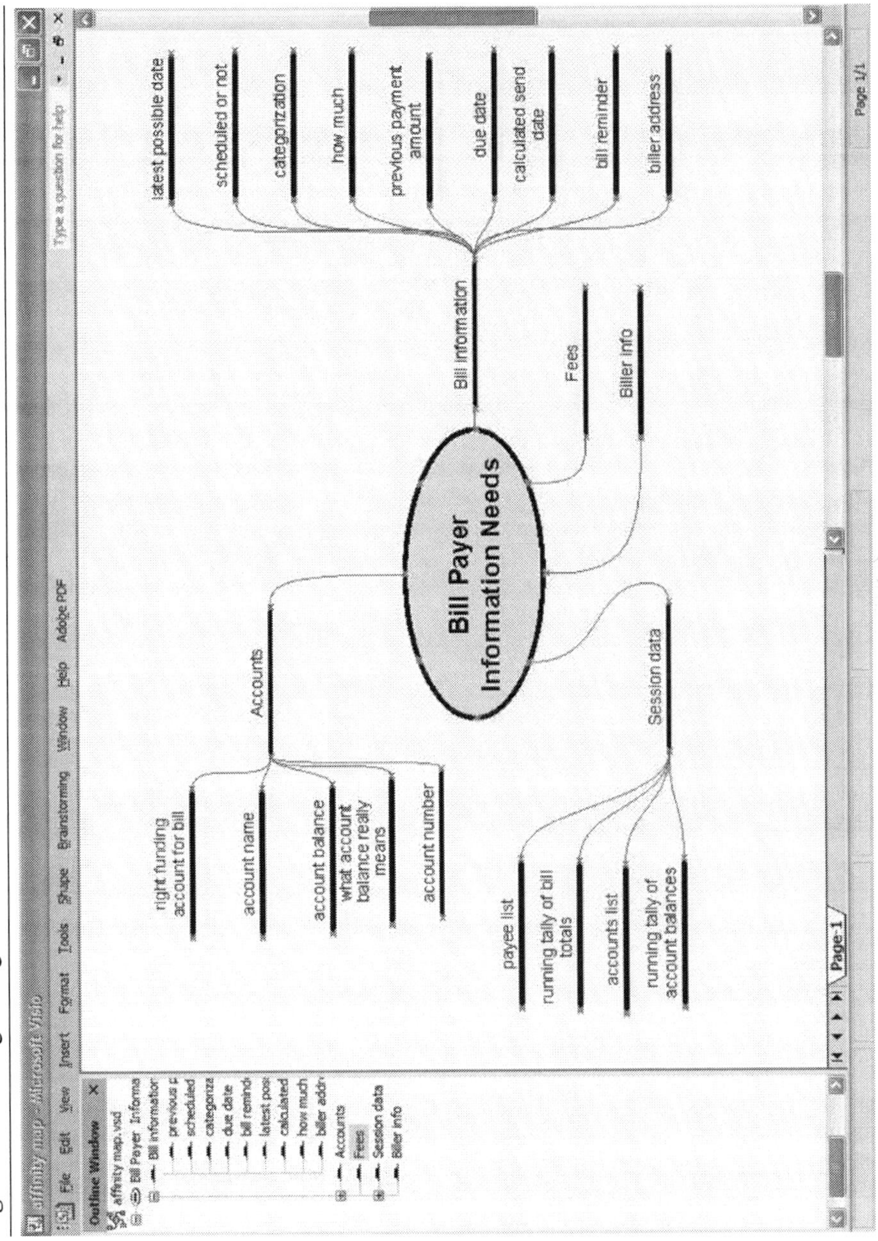

affinity map - Microsoft Visio
File Edit View Insert Format Tools Shape Brainstorming Window Help Adobe PDF
Type a question for help
Outline Window
affinity map.vsd
Bill Payer Informa
Bill information
previous p
scheduled
categoriza
due date
bill remind
latest poss
calculated
how much
biller addr
Accounts
Fees
Session data
Biller info

Bill Payer Information Needs
Accounts
right funding account for bill
account name
account balance
what account balance really means
account number
Session data
payee list
running tally of bill totals
accounts list
running tally of account balances
Bill information
latest possible date
scheduled or not
categorization
how much
previous payment amount
due date
calculated send date
bill reminder
biller address
Fees
Biller info
Page-1
Page 1/1

Figure 5.3 Example of a model shown as a table. (Hughes, M. 2000. Team usability testing of web-based training: the interplay of team learning and personal expertise. PhD diss., Univ. of Georgia. With permission.)

Facilitator Roles

	Moderator	Expert	Coach
Focus	Meeting/event efficiency	Fixing the product	Team and individual learning
Techniques	• Communicates time/procedural status • Directs	• Teachers principles • Puts findings into context • Recommends solutions	• Encourages participation • Encourages reflection • Encourages scrutiny
Knowledge Base	• Procedures, facilities, and equipment	• Design guidelines • Heuristics	• Group dynamics • Team learning • Action science
Data	User profiles/scenarios	Historical (other users in other tests)	Directly observable data within the user sessions

Figure 5.4 Example of a model shown as a process diagram. A pattern language approach to usability knowledge management. (From Hughes, M. 2006. *Journal of Usability Studies* 2 (1):76–90. With permission.)

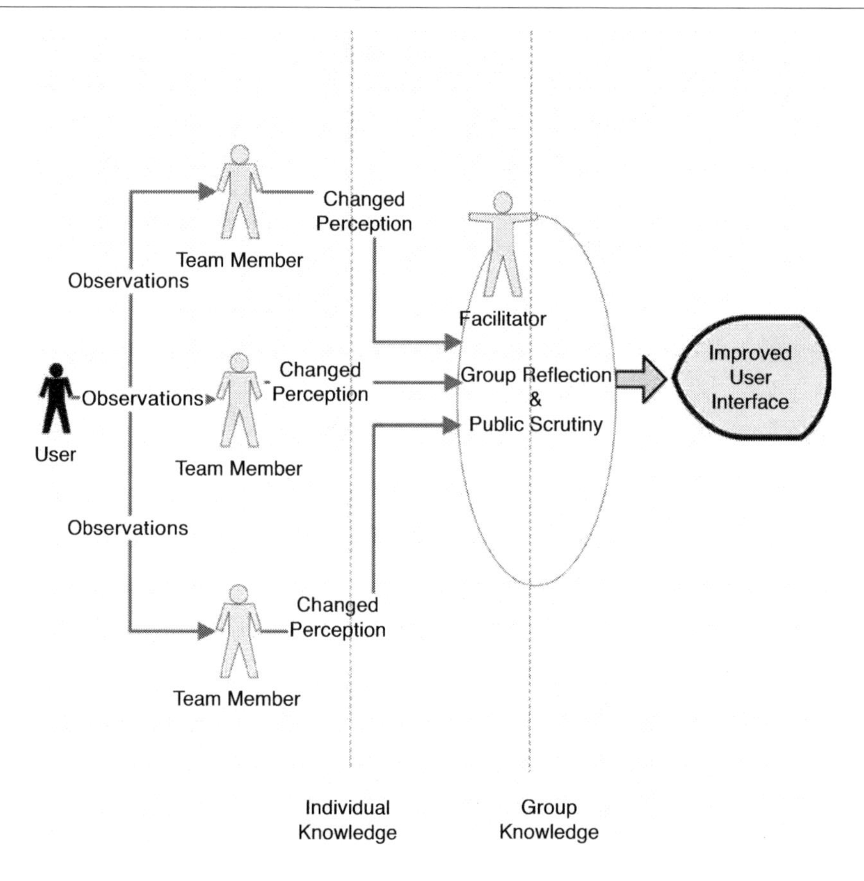

see direct quotes taken from the coded data to support principles or assertions made in the discussion section of the report. Not only does this approach help illuminate the conclusion, but it adds credibility.

Tools and Techniques

If you are contemplating a large research project, even a doctoral dissertation, purchasing specialized software for coding and categorizing qualitative data can be a useful investment. Several products can be evaluated using free downloads. (Many of these free demonstration packages are the full application with limited ability to save results.) An Internet search for "qualitative research software" will yield a good list.

But you can also take advantage of tools that are already available to you, namely, word processors such as Microsoft Word and graphing applications such as Microsoft Visio. Even spreadsheets can be useful for coding, categorizing, sorting, and filtering log sheets such as those created during usability tests. In this section, we discuss how to use Microsoft Word to analyze qualitative data.

Word Processor

If your data is in the form of an electronic text file, a word processor that has indexing capability can be an effective data analysis tool. Using Microsoft Word, for example, the overall process is as follows:

1. Open or import the documents or transcripts you wish to analyze into a Word document.
2. Go through the document using the indexing feature to code specific entries (see Figure 5.5).
3. Generate an Index (from the main menu, click Insert | Reference | Index and Tables).
4. Copy the index into a new document or a new section of the document you are analyzing.
5. Shift the View to Outline and initially select "Show All Levels."
6. Start creating categories and tagging them as Heading 3. Cut and paste (or click and drag) appropriate index entries under the categories you have created. Continue this step of creating categories and assigning index entries to those categories until all index entries have been assigned to a category.
7. Shift the View to "Show Level 3." Now you see just the new categories you created in Step 6.
8. Start creating a higher level of categories and tagging them as "Heading 2." Cut and paste the Heading 3 categories under the newly created Heading 2 categories until all the former categories have been assigned to the latter category.
9. If you want to continue categorizing, shift the View to "Show Level 2" and repeat the same process you used in Step 8 to create Heading 1 categories.
10. Finally, shift the View to "Show All Levels." You now have a list sorted by categories (and subcategories if you go to that level) with specific page references to the instances in the manuscript where each code was originally applied. See Figure 5.6 for an example showing an analysis of the first several pages of this chapter's manuscript.

Figure 5.5 Index dialog box.

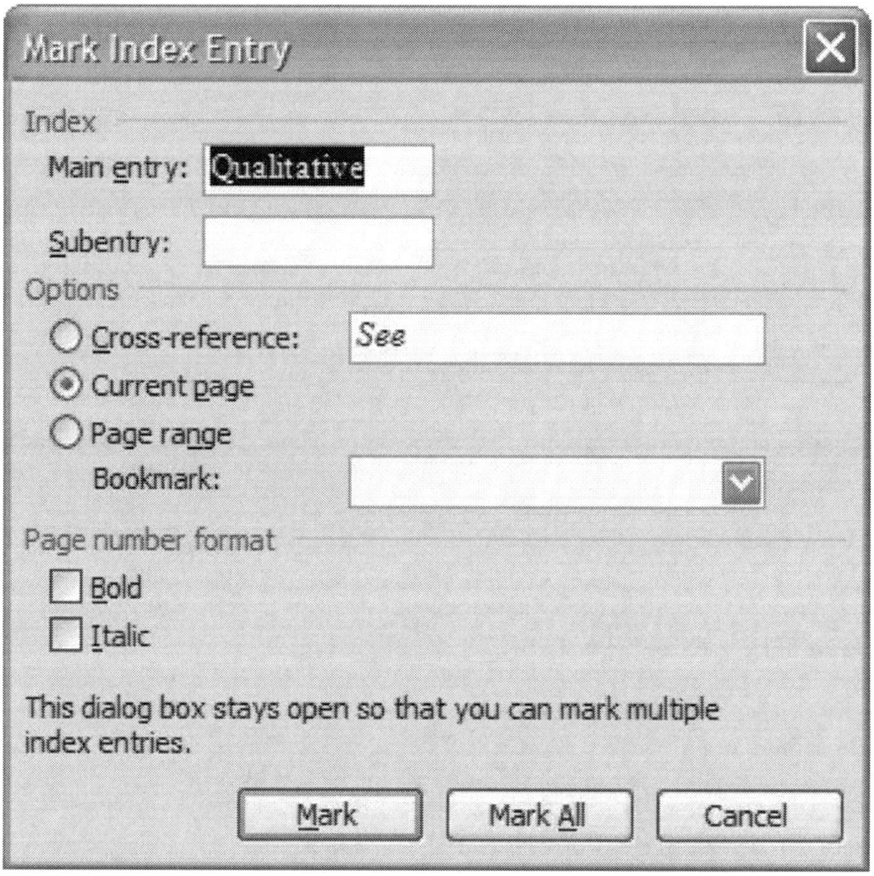

The resultant output is the findings of your analysis. The categories represent the themes that emerged from the data, and the index entries themselves are links back to the directly observed data (with the page numbers where the entries can be found in the original manuscript). This is essentially all that any qualitative data analysis tool can do. More specialized tools make it easier to analyze multiple documents and create visual models, but even for a master's-level research project, this simple word processor protocol can give robust results. The creative part of the process is what conclusions you draw from your results, what model you infer from the themes that emerge.

Summary

Qualitative data is nonnumeric—it consists of words, images, objects, or sounds that convey or contain meaning.

Qualitative methods generally involve the following three phases:

1. The observation of behaviors or artifacts in natural surroundings or authentic contexts
2. The recording or noting of those observations in descriptive or narrative formats

Figure 5.6 Index entries sorted into categories.

Standards

Credibility

credibility, 4, 5
internal validity, 4

Dependability

dependability, 8
external validity, 6

Transferability

transferability, 6, 7

Methods

Qualitative

qualitative methods, 2
document analyses, 3
field observations, 3
focus groups, 3
interviews, 3
usability tests, 3
analysis, 3
observation, 2
qualitative, 2, 4

Quantitative

quantitative, 1, 4

Society

Community

communities of practice, 1
community at large, 1
professional communities, 1

Social Studies

social implications, 1
social study, 1
sociology and anthropology, 1
technical communication, 1

3. The systematic analysis of those observations or notes to derive patterns, models, or principles that can be applied beyond the specific events or artifacts studied

Qualitative research has its own standards of rigor:

- Credibility: Do the participants truly represent the population or phenomenon of interest and how typical are their behavior and comments?
- Transferability: How natural or authentic was the environment in which the study took place?
- Dependability: How confident are you that the conclusions reached in the research project could be replicated with different researchers?

The following are common qualitative methods within technical communication research:

- Interviews—what respondents say when asked (one-on-one)
- Focus groups—what respondents say when asked (one-on-many)
- Usability tests—what users do when observed in a controlled environment
- Field observations—what users do when observed in their natural environment
- Document analyses—what documents tell us about the authors or the intended audience

Analyzing qualitative data consists of the following:

- Coding the data
- Categorizing the codes
- Drawing models out of the emergent categories and their relationships

References

Corbin, J., and Strauss, A. 1990. Grounded theory research: Procedures, canons, and evaluative criteria. *Qualitative Sociology* 13 (1):3–21.

Firestone, W.A. 1993. Alternative arguments for generalizing from data as applied to qualitative research. *Educational Researcher* 22 (4):16–23.

Hughes, M. 2000. Team usability testing of web-based training: the interplay of team learning and personal expertise. Ph.D. diss., University of Georgia.

———. 2006. A pattern language approach to usability knowledge management. *Journal of Usability Studies* 2 (1):76–90.

Kidder, L. 1982. Face validity from multiple perspectives. In *New Directions for Methodology of Social and Behavioral Science: Forms of Validity in Research,* ed. D. Brinberg and L. Kidder, 41–57. San Francisco, CA: Jossey-Bass.

Lincoln, Y.S., and Guba, E.G. 1985. *Naturalistic Inquiry.* Beverly Hills, CA: Sage.

Merriam, S.B. 1998 *Qualitative Research and Case Study Applications in Education.* San Francisco, CA: Jossey-Bass.

Seely-Brown, J. 1998. Research that reinvents the corporation. *Harvard Business Review on Knowledge Management.* Boston, MA: Harvard Business School Publishing.

6

Conducting Surveys

Introduction

Surveys are a popular way for student researchers and experienced researchers alike to gather data for their studies. With the emergence of free Internet surveying software and the proliferation of academic and professional list servers, surveying certainly seems to be an efficient way to reach a large sample and collect a relatively large body of data. But unless correctly designed, implemented, and analyzed, a survey can result in a botched opportunity at best or a misrepresentation of a population at worst.

Given the popularity of surveys as a research tool and their ability to support both quantitative and qualitative studies, we devote a chapter to them. As with the chapters on quantitative and qualitative methodologies, a single chapter cannot treat a topic as rich as surveying with the depth and scope needed to fully master it. Our objective instead is to give student researchers enough of an introduction to enable them conduct valid surveys and to understand what limitations might constrain their reliability. We also try to give the critical reader of research an understanding of the criteria and considerations that should be applied when reading research articles based on survey data.

Learning Objectives

After you have read this chapter, you should be able to do the following:

- Write clear, valid survey questions
- Calculate response rate, margin of error, and confidence interval

What Can Surveys Measure?

A survey is a list of questions that asks respondents to give information about specific characteristics, behavior, or attitudes. Surveys can help us understand the demographic profile of a sample or estimate a statistic within a population. For example, surveys can collect data about the respondents' age, gender, salary, years of experience, level of education, and so forth. These data points could be directly pertinent to the research question or could be used to assess how representative a sample is of the larger population from which it was drawn. For example, if a researcher were examining gender differences in salary levels among technical communicators, then asking the respondent's gender and salary would be directly related to the research question. If, on the other hand, the study was about college students' use of the Internet as a study aid, questions about age and gender might be used merely to determine whether the demographics of the sample were representative of the college's demographics in general.

Surveys can also measure behavior (or more accurately, self-reported behavior). For example, a survey can ask how frequently students conduct Internet searches in connection with a homework assignment.

Last, surveys are often used to assess attitude, that is, the respondents' feelings about a topic. Who among us has not taken a survey at one time or another that asked whether we "strongly agree" or "strongly disagree" with a statement such as "I would be likely to ..."?

Constructing a Survey

Constructing a survey consists of determining what questions to ask, asking them in the appropriate format, and then arranging them in an effective order. Additionally, you need to write instructions for completing and returning the survey.

What Questions to Ask

The questions you ask in your survey need to be driven by the research goal and research questions. Much of the art of surveying consists of constructing questions that operationalize the research goal and research questions appropriately. For example, a researcher might have the research goal of determining whether there is a correlation between writers' general use of the Internet and their receptiveness to online collaboration tools. That researcher must construct questions that determine how much time the respondent spends on the Internet, how often the respondent uses the Internet, or both. The researcher must also include questions that reveal attitudes about collaboration in general and online collaboration specifically.

There is no absolute formula for writing good questions, but the following guidelines can be useful:

- **Avoid absolute terms** such as *always* or *never.*
- **Avoid statements in the negative**. For example, consider this question: "Is the following sentence true for you? 'Subject matter experts are not a source of technical information for me.'" To say that subject matter experts *are* a source, the respondent must say "No" or "False." Using a double-negative to state a positive can lead to misunderstanding and thus to unreliable data. Therefore, word your questions so that "Yes" means yes and "No" means no.
- **Keep questions unbiased**. Consider the following question: "Do you feel that productivity-enhancing processes such as online editing improve your effectiveness?" This question certainly signals that the surveyor assumes that online editing enhances productivity. If the respondent does not share that perspective, this could be a difficult question for the respondent to answer honestly.
- **Focus on one concept per question**. "Double-barreled" questions combine multiple concepts and can lead to vague results. For example, the question, "To what extent are indexes and tables of contents useful to you in accessing data?" combines indexes and tables of contents in the same question. Some respondents might want to provide different answers for each. It would be better if that question were broken into two questions: one about indexes and another about tables of contents.

Question Formats

Survey questions can be categorized within the following formats:

- Open-ended
- Closed-ended
- Multiple-choice
- Rating
- Ranking

A well-designed survey uses the appropriate type question for the data it is seeking.

Open-Ended Questions

Open-ended questions allow free-form answers. The respondents can type or write sentences or even short essays to express their answers. Open-ended questions can give rich insight into respondent attitudes and can reveal unanticipated responses or themes. For these reasons, open-ended questions are particularly suited to qualitative research.

An example of an open-ended question is, what typically frustrates you the most when conducting research?

Some advantages of open-ended questions are the following:

- They give "voice" to the respondent; that is, they allow respondents to express their opinions in their own words. This deeper insight into the respondents' perspectives can enhance credibility in a qualitative study.
- They reduce the degree to which the researcher's framing of the problem is imposed on the respondent. In other words, they are less likely to bias the answer toward an expected response or outcome. This openness can increase the credibility and dependability of the results.

Some disadvantages of open-ended questions are the following:

- They are more demanding on respondents' time and require more energy to answer. This extra effort could reduce the overall response rate and detract from the reliability or dependability of the study.
- They are more demanding to analyze and report. If you expect to get 800 surveys back and are not prepared to analyze the qualitative responses, do not waste the respondents' time by asking open-ended questions.

Closed-Ended Questions

Closed-ended questions look for a single piece of data—a word or a number. They can be very useful for gathering accurate data, such as respondent age, salary, years of educations, etc.

An example of a closed-ended question is, "what percentage of your time do you spend on research?"

Some advantages of closed-ended questions are the following:

- They gather numerical data more accurately than multiple-choice questions (which typically use numeric ranges rather than specific numbers) would. This

precision is useful in quantitative studies, where statistical tests may require the standard deviation among the responses to certain questions.
- Verbal closed-ended answers are easier to categorize and summarize than open-ended answers.

Some disadvantages of closed-ended questions are the following:

- Verbal responses could be more varied and harder to analyze than if the question had presented a closed set of answers, such as a multiple choice.
- Questions seeking a numerical response might ask for a greater degree of accuracy than the respondent is able to give: for example, "How many employees are in your company?" (see "Multiple-Choice Questions" below for a better way to present that question).

Multiple-Choice Questions

Multiple-choice questions provide an array of choices from which the respondent selects one or more answers (ensure that it is clear whether more than one answer is allowed). If the survey is being administered electronically, use radio buttons for the choices where only one response is allowed and use checkboxes if more than one choice is allowed.

Figure 6.1 shows an example of a multiple-choice question where only one choice is allowed. Note that all choices are mutually exclusive. For example, the choices are not 100–500 and 500–1000. (Which would be the correct selection if the respondent's company had 500 employees?) Also note that the list is exhaustive. What if it had stopped at 1000–5000 and the respondent's company had 6500 employees?

Figure 6.2 shows an example of a multiple-choice question where more than one answer is allowed. Not only is that fact implied by the use of check boxes, it is explicitly stated in the question.

Figure 6.1 Sample of a multiple-choice question where only one selection is allowed.

What is the size of your company? (number of employees)

○ less than 100
○ 100-499
○ 500-999
○ 1000-5000
○ greater than 5000

Figure 6.2 Sample of a multiple-choice question where multiple selections are allowed.

How do you search for information in a document? (check all that apply)

☐ table of contents
☐ index
☐ electronic word search
☐ scan the headers and footers
☐ browse looking at headings
☐ browse looking at illustrations

Some advantages of multiple-choice questions are the following:

- They are easy for respondents to answer.
- They are easy for researchers to summarize.
- They can be analyzed automatically by survey software applications or optical-scanning devices.

Some disadvantages of multiple-choice questions are the following:

- They can leave out meaningful choices that the researcher did not anticipate. For example, in Figure 6.2, are there other ways a reader could look for information that are not listed? This problem can be mitigated by allowing a choice called "Other" and letting the user provide additional choices. But in that case, the researcher gives up some of the convenience of a multiple-choice and takes on some of the inconvenience of an open-ended or closed-ended question.
- They force the researcher's frame of reference on the respondent. The respondent's choices are constrained by the researcher's view of the question and, in fact, none of the answers might be acceptable to the respondent.

Rating Questions

A rating question is a specialized form of question that asks respondents to choose from answers that express, for example, degree of agreement such as with a stated opinion or a frequency to a question that asks, "How often do you...," along an axis. Different types of scales can be used for ratings.

A *Likert* scale presents a statement to which respondents indicate their degree of agreement or disagreement, typically along a five-choice scale. Figure 6.3 is an example of a rating using a Likert scale.

The same question could have been presented in a closed-ended format: On a scale of 1 to 5, how strongly do you agree or disagree with this statement: "Technical communicators do not get the same respect as other professionals in my company" (1 = strongly agree; 5 = strongly disagree).

A *frequency* scale asks respondents to indicate a frequency of a behavior. Figure 6.4 shows an example of a frequency scale.

Figure 6.3 Rating question using a Likert scale.

How strongly do you agree or disagree with this statement:

"Technical communicators do not get the same respect as other professionals in my company."

○ strongly agree
○ somewhat agree
○ neutral
○ somewhat disagree
○ strongly disagree

Figure 6.4 Frequency rating.

How often do you include indexes in the publications you produce?

- ○ almost always
- ○ often
- ○ sometimes
- ○ seldom
- ○ almost never

Figure 6.5 Semantic differential rating.

Please rate the Acme Content Management System among the following criteria (circle your answers):

Hard to learn	−3	−2	−1	0	+1	+2	+3	Easy to learn
Hard to use	−3	−2	−1	0	+1	+2	+3	Easy to use
Poor search capabilities	−3	−2	−1	0	+1	+2	+3	Good search capabilities

A semantic differential scale usually measures a series of attitudes toward a single concept, typically using a 7-point response scale with bipolar descriptions at each end. Figure 6.5 shows a semantic differential rating.

Some advantages of rating questions are the following:

- They can help clarify respondents' attitudes and preferences.
- They are easy for respondents to answer.
- They are easy for researchers to summarize.

Some disadvantages of rating questions are the following:

- Repetitive use can result in the respondent giving the same rating by force of habit.
- Numerical scales could be misunderstood by the respondent; that is, a respondent might use 5 to mean a top rating when the researcher intended 1 as the top rating.

Ranking Questions

Ranking questions ask respondents to order items along a defined dimension, such as importance or desirability. Figure 6.6 is an example of a ranking.

An advantage of ranking questions is that they can assess the relative priorities of different attributes.

Some disadvantages of ranking questions are the following:

- Respondents can easily confuse ranking questions with rating questions.
- Answers are not to scale. For example, for one respondent the first and second items might be critically important and the other items on the list not at all important, whereas for another respondent the importance might be distributed more proportionally. So a "3" from one respondent does not correlate to a "3" from another—at least as a measure of importance.

Figure 6.6 Ranking question.

Rank the following from 1 to 5 in the order of their helpfulness to you as a resource (1 is the most helpful).

_____ subject matter experts

_____ current documentation on previous version of product

_____ marketing collaterals

_____ manager

_____ fellow writers

Survey Structure

At its beginning (or in an accompanying cover letter/e-mail), a survey should explain the purpose of the study. It should also explain how the privacy of the respondents are being protected and that participation is voluntary.

The survey instructions should indicate the approximate amount of time it will take the respondents to complete the survey. If the respondents cannot stop and save their responses to complete the survey later, that fact needs to be made clear.

The survey needs to provide all required instructions or clarifications. Do not overdo the explanations—you do not want the respondents to get bored and abandon the survey. But ensure that you give all needed information.

Arrange the questions into logical groupings. If the survey is long, divide these groupings into sections to give the respondents a sense that they are making progress through the survey. At the end of each section of a long survey, a percentage-complete indicator will reinforce that sense of progress.

See the section in this chapter on response rate for an explanation of how the order in which questions are presented can influence the response rate.

End the survey with instructions about how to return or submit the survey, and thank the respondents for their participation.

Reporting Survey Results

Depending on the questions asked (quantitative or qualitative), the data analysis methods and standards of rigor described in Chapter 4 and Chapter 5 also apply to surveys. For example, if a survey collects numerical data about respondent salary, gender, and years of education, the researcher could use a *t*-test of two means to see whether there was a statistically significant difference between the salaries of men and women, or the researcher could determine whether there was a correlation between salary and years of education. If the survey contained open-ended questions that asked respondents to describe how they felt they were perceived within their organizations, qualitative coding and categorization methods described in Chapter 5 could be applied to look for themes or models that emerged from their answers.

A method of reporting results that we have not discussed yet, and one that is common in surveys, is *frequency distribution*.

Figure 6.7 Frequency distribution.

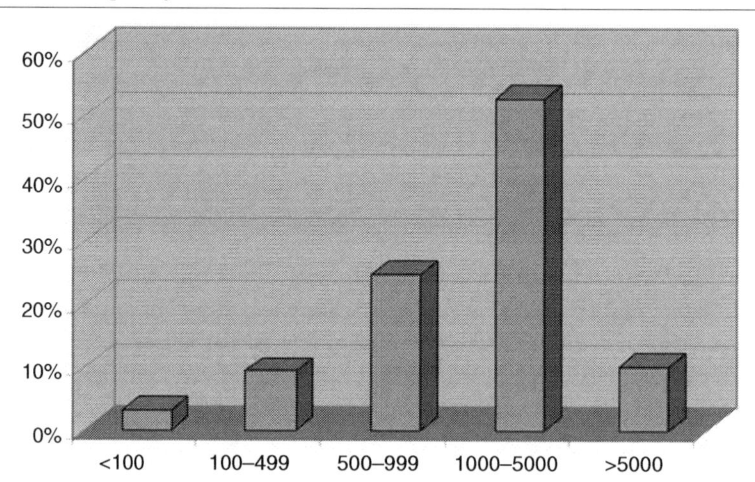

Frequency Distribution

A frequency distribution is a useful way to report how respondents answered multiple-choice questions, and the results are generally expressed as percentages. For example, respondent answers to the question in Figure 6.1 (size of company) could be summarized as in Figure 6.7. Frequency distributions can also be reported in tables. Each question of import could have its own frequency distribution, or related groups of questions could be combined into one frequency distribution (assuming they all had the same choices).

Measures of Rigor

Quantitative surveys are subject to the same measures of rigor discussed in Chapter 4, and they can also require reporting or analyses not discussed yet—specifically, response rate, margin of error, and confidence intervals.

Response Rate

Survey reports often include the response rate. To calculate the response rate for a survey, divide the number of surveys completed by the total number you sent out. For example, if you put a survey out on a list server that had 1000 members and you got 375 back, your response rate would be 0.375 or 37.5%.

The response rate gives an indication of how representative the returned surveys are likely to be of the population surveyed. For example, if a professional association sends out a salary survey and gets a response rate of 5%, the reader of a report based on that survey's data should be cautious about how representative the results are. A factor that can compromise the reliability of any survey is that the respondents self-select whether or not to return the survey. This self-selection aspect undermines the principle of random or systematic sampling—an important consideration if statistical tests or analyses are being applied to the results. The higher the response rate, the more likely that all segments of the population being surveyed are being proportionately represented.

There is no hard rule for acceptable response rates, but the Division of Instructional Innovation and Assessment at the University of Texas posts the following guidelines on their Web site (http://www.utexas.edu/academic/diia):

Mail: 50%, adequate; 60%, good; 70%, very good
Phone: 80%, good
E-mail: 40%, average; 50%, good; 60%, very good
Online: 30%, average
Classroom paper: ≥50%, good
Face-to-face: 80–85%, good

There are steps you can take as the survey designer and administrator to improve response rates:

Design a usable survey—If respondents struggle to understand how to answer or submit the survey, they are likely to abandon it.

Keep the length manageable—Respondents get tired or can grow impatient if the survey is too long; ask only what you need to.

Provide incentives—Some surveys offer cash rewards for participation; some put the names of the respondents into a drawing for a prize. Sometimes, just offering to share the results with the respondents can be enough of an incentive for them to complete the survey.

Note: Teachers who are surveying students or managers who are surveying employees need to be sensitive of the ethical concerns of coerced participation in research.

Arrange the questions to pique the respondent's interest—Do not start with the dull, demographic questions, such as age, gender, etc. Start with questions that are relevant to the goals of the survey and that are likely to be of interest to the respondent. However, if some questions are likely to be sensitive, wait until you have earned the respondent's trust before asking them. For example, if a researcher is conducting a survey to assess technical communicators' ethical practices, the question "Have you ever lied to a client or employer?" might be a useful question, but not one that should be asked at the beginning of the survey.

Margin of Error

When responses are reported as frequency distributions and the researcher wants to infer that the results from the survey sample are representative of the population of interest (which is usually the case in research), then the report should include the *margin of error*. The margin of error indicates the reliability of the inference and can be reliably estimated using just two inputs: the size of the sample and the required level of confidence. The margin of error is a value, in percentage points, that essentially lets the researcher say, "I think the real value in the population falls within the reported value plus or minus this margin of error." The confidence level is an indication of how reliable that statement is. Confidence levels of 90, 95, and 99% are common. So, if a researcher states that a survey has a margin of error of ±3 at a 95% confidence level, that statement means, "The true frequency percentage for 95% of the samples I could have taken is within ±3 of this number." In other words, the sample the researcher actually took could be part of the 5% that "misses" the true value. Note in Table 6.1 that the higher

Table 6.1 Margins of error (as a percentage)

	Confidence level		
Sample size	**90%**	**95%**	**99%**
25	16	20	26
50	12	14	18
100	8	10	13
500	4	4	6
800	3	3	5
1000	3	3	4
2000	2	2	3

the confidence level for a given sample size, the wider the margin of error. The analogy is: The bigger the net you cast, the better the odds of catching a fish; so the wider the margin is, the better odds you have that it includes the true value in the population. The trade-off is that the more confident you want to be in your estimate, the less precise the estimate will be; conversely, the more precise your estimate is, the less confidence you can have in it.

The good news is that a close approximation of the margin of error is easily calculated (or you can use Table 6.1 to get an idea of what the margins are at various sample sizes and levels of confidence). The formula is

$$\text{Margin of error (in percentage points)} \approx x/\sqrt{n}$$

where $x = 82$ for 90% confidence, 98 for 95% confidence, or 129 for 99% confidence, and where $\sqrt{n}$ = the square root of the sample size.

The bad news is that most researchers (especially student researchers) are disappointed when they see how large the margin of error is for their sample size. For example, let us say a researcher surveys 25 classmates and 40% respond that they regularly include indexes in their documents. Using a confidence level of 95% and referring to Table 6.1, the researcher sees that the actual percentage in the population could be 40% plus or minus 20, that is, between 20 and 60%. That is a wide margin of error. The researcher has two options:

- Increase the survey size (but he or she would have to get about 800 surveys back to be able to claim a respectable margin of error of ±3).
- Go ahead with the report (as a student researcher would almost certainly have to do) and report the margin of error.

Unfortunately, many researchers merely report the findings and talk about them as if they were representative of the population as a whole. As a critical reader of research, you can protect yourself by calculating the margin of error yourself, as long as the researcher has reported the sample size.

Confidence Intervals

Some surveys ask closed-ended quantitative questions, such as years of experience, salary, and so on, and therefore can report an average for the sample. If the researcher wants to infer that the sample average is representative of the population (which

Figure 6.8 Confidence calculation in Microsoft Excel.

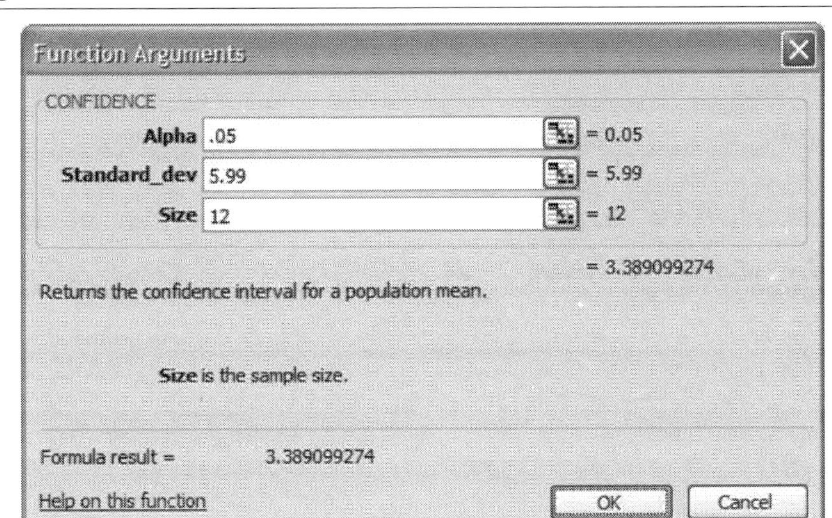

researchers typically want to do), then the researcher should indicate the reliability of that inference by reporting the *confidence interval*. Just as with the margin of error for frequency distributions, the confidence interval is a range within which the researcher is confident (once again, typically at 90, 95, or 99% levels) that the actual value falls.

When giving the confidence interval for a mean (average), the calculation is based on sample size (*n*) and on the standard deviation (*SD*). In Chapter 4, we show how to use a spreadsheet to calculate *n*, mean, and the *SD*. The Microsoft Excel function CONFIDENCE can be used to derive the confidence interval for that mean, based on the *SD* and the confidence level you want.

For example, let us say that a researcher surveys 12 technical writers and asks for the number of years experience each has. Using the methods shown in Chapter 4 to determine mean and standard deviation, the researcher calculates the mean experience to be 7.08 years and the standard deviation to be 5.99. The researcher could insert the function CONFIDENCE into the spreadsheet and would be presented with a dialog box like the one shown in Figure 6.8. In our example, the researcher wants a confidence level of 95% so the researcher puts in 0.05 for Alpha. Alpha is 1 minus the confidence level (expressed as a decimal) you want; in our example, 1 minus 0.95 equals 0.05. The researcher then enters the standard deviation (5.99) and the sample size (12). The result is 3.389. The confidence interval, then, is the mean plus and minus this calculated value.

If the researcher rounded the years of experience to 7 and the confidence result to 3.4, he or she could report, "We can be 95% confident that the average experience level for a technical writer is between 3.6 and 10.4 years." As with the margin of error discussed earlier, small sample sizes result in wide confidence intervals. As with margins of error, the researcher should report the confidence interval so the reader can determine how much weight to place on the finding.

Unfortunately, many researchers fail to do—or report—this calculation, and they talk about the estimated statistic as if it were a reliable inference about the population as a whole. As a critical reader of research, however, you can easily perform the confidence calculation for yourself as long as the researcher reports the sample size and the standard deviation.

Table 6.2 Survey exercise

Surveys sent	Surveys returned	Response rate	Margin of error (confidence level = 95%)
95	32		
1500	325		
1500	872		

Exercise 6.1: Reporting survey results

1. For each of the studies in Table 6.2 provide the response rate and the margin of error.
2. You survey 32 students in a technical communication master's program and ask how much time elapsed between completing their bachelor's degree and enrolling in a master's program. You would like to use this data to make an inference about how long technical communicators typically wait before pursuing a master's degree. The average of the responses was 7.5 years and the standard deviation was 4.5. If you wish to have a 90% confidence level, what are you willing to say about the average time technical communicators wait between getting a bachelor's degree and enrolling in a master's degree program? ■

Answer Key

Exercise 6.1, Question 1

Surveys sent	Surveys returned	Response rate	Margin of error
95	32	34%	±17
1500	325	22%	±5
1500	872	58%	±3

Exercise 6.1, Question 2

"We can be 90% confident that the average time between getting a bachelor's degree and enrolling in a master's program is between 6.2 and 8.3 years." **Note:** You should have used the CONFIDENCE function in Microsoft Excel with an Alpha of 0.1 (as the confidence level is 90%), a Standard_dev of 4.5, and a Size of 32. This results in a confidence interval of 7.5 years plus or minus 1.3 years.

Part Two

EXEMPLARS AND ANALYSES

7

Analyzing a Literature Review

Introduction

In Chapter 3, we explored literature reviews in terms of the roles they play in primary and secondary research, the purposes for preparing them, the skills they require of the researcher, and a methodology for writing them. In this chapter, we will analyze a specific literature review to see how one author has approached the task of exploring previous work on his topic. The chapter contains the full text of Peter MacKay's "Establishing a corporate style guide: A bibliographic essay," which appeared in *Technical Communication* in 1997, as well as a detailed commentary.

Learning Objectives

After you have read this chapter, you should be able to

- Analyze a literature review in terms of purpose and audience, organization, citations and references, and level of detail
- Apply your knowledge of these analytical techniques to preparing a review of the literature on your topic of interest

The Article's Context

As we noted in Chapter 3, few literature reviews (sometimes called bibliographic essays) or annotated bibliographies are published in the technical communication journals. Much of the reason for the scarcity of such articles seems to be the stigma that some academics attach to what they consider "derivative research," even though an effective literature review or annotated bibliography makes an important contribution to the body of knowledge in our field by collecting citations of essential works on a topic, describing their content, and sharing insights about their significance. Although such articles do not constitute primary research, they provide a valuable service to other researchers.

The article we examine in this chapter does just that. Peter D. MacKay examines in detail 14 articles and conference papers published over a 10-year span, classifies them under five perspectives of the task of preparing a style guide, and draws an important conclusion about the state of research on this important topic. Although the article is now a decade old, relatively few articles and conference papers have been published on the topic since it appeared, so it remains surprisingly current and continues to provide important insights into corporate style guides.

"Establishing a Corporate Style Guide: A Bibliographic Essay"*

Peter D. MacKay

Deciding whether to establish a house style guide can be a difficult decision for writing departments. Management must decide whether it is worth the time, money, and energy to develop its own specialized style guide when various general style books already exist on the market. And if a company does decide to go ahead and establish a house style guide, what form should the document take? Will the guide be effective? What considerations should be weighed in determining whether house style rules should be established in a particular business? Published studies in this area are somewhat sparse, but several recent articles and conference presentations can help answer these questions. They do so by bringing important factors to our attention, ranging from explanations of why and how a style guide can improve the quality and consistency of a company's documents to the strategies for developing the guide itself.

For people who work in the field of written communication, questions of style pose ever-present concerns. Practitioners strive for documents of high quality and are naturally concerned with issues such as whether to use active or passive voice, capitalize titles, abbreviate terms, hyphenate or close up words such as *free-lancer*, use one "l" or two in a word such as *cancelled*, and use footnotes or in-text citation methods. Practitioners also must frequently balance the need to eliminate gender bias and unnecessary jargon in prose, while retaining certain aspects of field-specific terminology. And these are only a few possibilities from a long list of other issues that can arise on a regular basis. We frequently encounter different writing styles in casual reading. Perhaps a reader notices that writers for *The New York Times* use *Ms. Jones* or *Dr. Wilkes*, whereas the authors of articles in many other publications simply write *Jones* and *Wilkes*. Style issues sometimes seem like preference choices—one writer or publication may opt to do something differently from another. Yet organizations usually want to set ground rules, so they decide to use a consistent style. The assumption is that a consistent form adds to the credibility of a company's publications, whereas inconsistencies detract from the company's image. Thus, the issue is an important one, whether a writing department consists of 1 person or 100 people.

Though the notion of what precisely constitutes a "style guide" may vary from setting to setting, a basic framework for discussion should be agreed on. I propose the following as a basic working definition: *A style guide is a rule-driven document that sets the parameters for consistency and acceptability for all written materials produced by an individual or group. A house style guide is one that is produced for an organization's internal use and is specifically tailored for its specific writing contexts.*

This bibliographic essay presents an overview of recent research on style guides, critically analyzes these studies, and suggests areas for future consideration. The

---

* This article was originally published in *Technical Communication* 43:244–251. References' and citations' styles have been changed to conform to *The Chicago Manual of Style*, 15th edition. Reprinted with the permission of the Society for Technical Communication.

articles are grouped in thematically and analytically related areas. That is, the articles can be grouped in terms of emphasis on certain topics and the different types of analytic treatment the authors give to these topics. Of course, these groupings are somewhat arbitrary, and there is some overlap between these areas. I hope, however, that reviewing the material on the basis of related general thematic and analytic treatment will help professionals who want to know what has been written on issues relating to style guide development. This review is limited to work published between 1985 and 1995, a limitation that isolates the focus to reasonably current work, while providing a significant time frame for a range of research projects.

Planning Development: Determining Form and Content of a Style Guide

One thematically and analytically related group of articles focuses on establishing a plan for the development of a house style guide and determining the manual's basic form and content. Perhaps the most straightforward set of options is presented by Sharon Trujillo Lalla (1988). In "The state-of-the-art style guide development," Lalla deals directly with the question of whether an organization should adopt a standard professional style guide, create a house style guide, or combine both of these options. The answer, she suggests, depends on the situation. Among the key areas to consider are organizational needs and the level of management support. A preliminary analysis of existing organizational documents is necessary, followed by a memo to management outlining suggested courses of action. Lalla emphasizes that the style guide developer, through the preliminary analysis, must assess management's commitment to the project. Without strong management support, efforts to establish a style guide will not succeed. However, this argument operates on the assumption that the style guide developer and management are different people.

After conducting the preliminary analysis and obtaining management reaction, the style guide developer can determine which strategy to take. Lalla says that a "standard" style guide is a good option if management support is minimal or if a general guide from a particular discipline can be applied. This approach is straightforward, involving review of various existing style guides and choosing one that seems to fit well with the organization's needs. Creating a house style guide is, Lalla contends, a more comprehensive task. It focuses on problems unique to the organization and may involve creating guidelines from scratch. She suggests that it is necessary to form a committee, make lists of common writing problems, meet regularly, assign tasks or select discussion topics prior to meetings, and make decisions on the production of the document itself (177-178). The final option, using both a standard style guide and a house style guide, may also be an attractive alternative, depending on the situation. A company, she suggests, may want to use a standard style guide for general stylistic issues, but adopt a house style guide for more specific writing concerns that occur frequently in the types of documents the company produces.

One advantage of this approach is that it may save some time and effort in producing a guide, but, as Lalla points out, a disadvantage is that users may be frustrated in having to consult two guides to find the answers to their questions. Lalla says that whatever approach is taken, it is necessary to conduct a review of the guidelines after drafts are distributed, announce that the style guide represents organization policy, and devise a revision procedure for issues that may later arise. Lalla's options seem to be geared toward a large company, with a clearly delineated hierarchical structure and an active staff from whom answers can be easily obtained. But what approach should he taken if a manager is the style guide developer, if staff does not actively participate, or if the writing department consists of a single person?

In "A style guide—Why and what," Valerie Mitchell (1986) deals more with the content of style guides than with the dynamics of producing one. She says that a good style guide provides preferable word choices, determines what jargon is acceptable, sets rules to follow, and chooses authoritative dictionaries and reference sources. She, like Lalla, contends that a style guide should be comprehensive. Mitchell also suggests that it should deal with all aspects of developing publications. A style guide, she says, makes writers aware of their responsibilities and "ensures adherence to company standards, and at least *some* semblance of format" (232). Although Mitchell says that a style guide needs to promote only some semblance of format, other authors would hold that format should be a major concern. Mitchell concludes her brief article by providing a list of possible topics and subtopics for a style guide, including not only writing issues but also considerations of planning, scheduling, and formatting technical reports.

In "The evolution of a style guide," Clifford M. Caruthers (1986) traces the development of *Computing Services Writing and Editing Standards,* a 100-page internal style guide for Argonne National Laboratory Computing Services. The project, he notes, grew out of a request from the organization's director for consistent naming, usage, and punctuation conventions. According to Caruthers, a set of prescriptive rules was originally developed to deal with these areas, but within a year, the organization decided that this style guide was incomplete; these rules were later incorporated as a chapter within the updated style guide, while new chapters dealt with references to useful sources, suggestions for improving writing style, communicating news in an effective manner, and publishing technical information.

Unlike Lalla and Mitchell, Caruthers believes that it is not necessary to develop a comprehensive style guide. Argonne National Laboratory Computing Services did not want a guide that overlapped with style guides already available on the market. "What we needed," he writes, "was a guide that would complement these texts by providing our authors—experienced professionals in computing services but mostly untrained as technical writers—with some key advice on style and tone." The house style guide would deal with specific situations that may arise with in-house documents (for example, using "Job Control Language" rather than the acronym "JCL" and permitting common terms such as "Fortran" without further detail), but would suggest William Strunk and E. B. White's *The Elements of Style* and other guides for general prose concerns. Thus, the house style guide would be detailed, but only to a certain point and not to the extent of including information covered elsewhere.

The articles by Lalla, Mitchell, and Caruthers offer different views on planning for style guide development and determining a publication's form and content. Lalla presents three options for style guides, while Mitchell and Caruthers have narrower views of what a style guide should constitute. Lalla mentions that the situation itself is the overriding factor in determining what style guide is appropriate for a given organization, whereas Caruthers suggests through his narrative that situation may be important.

Time, Financial, and Management Issues

In addition to outlining goals for development and considering options for the possible form and content a house style guide might take, a style guide developer needs to consider whether the project is worthwhile before embarking on the endeavor. A few authors have pondered time, financial, and management factors—issues dealing with the project's feasibility from an organizational perspective.

In "The nine-year gestation of a unified technical style guide," J. Paul Blakely and Anne S. Travis (1987) outline the process and time required to develop a style guide for a scientific and engineering organization with 17,000 employees in four locations. Each plant had basic style guides, and manuals also existed for departments and divisions within departments at these plants; Blakely and Travis estimate that there were once 150 to 200 style guides in the company. The organization, Union Carbide, decided to consolidate its operations, and a committee was formed to develop a single, unified style guide. According to Blakely and Travis, because members of the committee had different professional backgrounds and interests, conflicts sometimes occurred, so a system of compromise and agreement had to be developed to keep the group focused. The authors note that the group's chairperson developed schedules and gave assignments to the members of the committee, and members had to report on assigned action items at weekly meetings. The group then debated various issues, arrived at decisions by voting, and presented proposals to upper management for approval, often a chapter at a time.

The development process, Blakely and Travis explain, was a lengthy one. After several chapters were completed, drafts were distributed, and readers had the opportunity to provide their reactions on comment forms; the authors state that it took nearly two years to resolve all the comments and reach a final draft. Also, along the way, funding problems arose; the authors note that the project was nearly completed when funds ran out, and work stopped for about a year until a grant was received. From inception to completion, the entire development process lasted a total of nine years, from 1977 to 1986. This presentation demonstrates that developing a style guide can be a monumental task.

For such a task, as Blakely and Travis mention, strong leadership is important. Producing a style guide can become a management issue for several reasons. In "Aren't you glad you have a style guide? Don't you wish everybody did?," Patricia Caernarven-Smith (1991) addresses the management perspective. She reasons that:

- Managers have the responsibility for every publication that leaves their departments.
- Managers assign projects to writers and editors in their departments, and everyone in the chain needs to understand the appropriate style.
- Management must maintain a budget for style guides and other resources.
- Management needs to be aware that engineers and programmers may try to influence the writing and editing process, and style guides can help keep matters clear.
- Management needs to reduce the time spent discussing problems that a style guide could solve.

Likewise, in "Finding solid ground: Using and articulating the grammar of technical editing," Victor W. Chapman and Jean L. Owens (1986) discuss "midwifing" a style guide, again highlighting the managerial and leadership roles. Chapman and Owens describe a process of holding weekly meetings, appointing a neutral moderator, encouraging members to research and discuss upcoming topics, ensuring that everyone has an opportunity to speak, keeping meetings focused, conducting votes, saving notes for future reference, and initiating a yearly review of the style guide after its publication.

Neither Chapman and Owens's study nor Caernarven-Smith's article deals directly with the cost involved in producing a style guide. Both reports focus on the overall management picture, presenting editorial arguments about how a house style guide can make the process of document production a smooth, efficient task. Perhaps one reason for not mentioning the cost of developing a style guide is that this factor will vary considerably from one organization to another. And even if potential style guide developers possess the best leadership and managerial skills, Blakely and Travis's description of funding problems may be enough to scare them away.

Others take a different view of the financial aspects involved. Paul R. Allen, in "Save money with a corporate style guide" (1995), contends that style guides "provide big dividends for a small investment," saving money in the long run (284). He says the reasons why an organization should develop a house style guide include creating consistency in documents, promoting a professional image, training new employees, and guiding document generation. "A common thread among these reasons to develop a corporate style guide," he claims, "is to save money, which is the equivalent of making money" (285). The money issue, Allen notes, is often not discussed in articles on style guide development, though it sometimes can he inferred from other issues. For example, discussions of saving time or eliminating disagreements between writers and editors often translate into financial issues since "time is money" in the business world. And when authors do mention the topic directly, they usually devote little time to it. Regardless of how an organization states its goals, however, the bottom line is that a company aims to make money, or at least recover what it invests. Allen suggests that reducing costs is the predominant reason why corporations should develop style guides.

Readers might expect Allen to provide hard evidence to show how style guides have saved money for a lot of corporations—a dollars-and-cents comparison, statistics of projects that were over and under budget, or some type of empirical data indicating how money was saved or spent. But Allen does not provide such supporting data. Instead, he does exactly what he criticizes others for doing: highlighting the various advantages of a style guide and outlining the steps to development, while neglecting to discuss exactly how the document saves money for the company. For example, he describes how a style guide can provide consistency in documents for a business with many writers or a company with offices in several distant locations; breakdowns in communication can occur in these scenarios, but a style guide can help bridge the gaps. On the topic of consistency, Allen reiterates Caernarven-Smith's point that the number of disputes among writers and editors can be reduced if a style guide is in place. He also echoes others in his contention that a style guide can help avoid problems of answering style questions "on the fly." Allen concludes the section on consistency with the statement that "a corporate style guide saves time, which saves money" (286), but readers are left wondering about specific details, including any estimates of the amount of time or money saved. Allen takes much the same approach in his sections dealing with how style guides can help promote a professional image of a company, train new employees, and define document generation.

Allen then shifts gears, discussing how a corporate style guide should be developed. At times, he raises some very specific points. For example, he notes that it is a good idea to put the document in a three-ring binder, rather than having it bound, so updates can be added without reprinting the entire guide (288). He also suggests including a comment form in the back of the document, so users can easily complete and submit their comments for update considerations (288–289). Again, there may be some implied relationships to the financial aspects of creating and revising a style guide (for example, updates to the guide can be made in a relatively inexpensive manner), but Allen shies away from using any specific figures to drive these points home.

Steps for Developing a House Style Guide

After considering the options available and the project's feasibility, a style guide developer will want to decide exactly what strategy to take in creating a house style guide. Some authors have suggested steps, often in checklist format, for developers to consider. In outlining steps for developing a house style guide, Durthy A. Washington is a clear frontrunner among the authors I have considered so far. Her work is cited frequently in recent studies on style guides. Two articles by Washington are important in this area: "Developing a corporate style guide: Pitfalls and panaceas" (1991) and "Creating the corporate style guide: Process and product" (1993).

In her 1991 article, Washington identifies four steps to developing a style guide:

1. Clearly defining project goals and objectives
2. Gathering information from prospective users
3. Deciding what format the guide should take
4. Testing the final product

Gathering information can be done formally or informally through meetings, interviews, and surveys. Testing the final product should involve getting reactions, through a variety of techniques, from users of the style guide. The objective, according to Washington, is to determine whether the guide is clear (with a clean and uncluttered format), comprehensive (with accurate and up-to-date information), easy to use (with a thorough index, table of contents, and appendices), and attractive (that is, "packaged" nicely to give a positive visual image).

Washington's 1993 article covers much of the same ground as the earlier paper, including some passages directly extracted from the earlier work. However, in this follow-up article, Washington identifies more steps, and even sublevels of steps in some cases:

1. Conducting a requirements analysis
2. Defining the project scope

3. Organizing the project
4. Writing the product description
5. Creating the mockup
6. Incorporating design elements
7. Developing the draft
8. Evaluating the final guide
9. Maintaining the style guide
10. Promoting the style guide

Washington describes the steps in a straightforward, common-sense tone, and gives the first and last steps the most extensive treatment. In conducting a requirements analysis, Washington suggests that one needs to assess the available resources, review current documents, identify baseline problems, and solicit management support. In promoting the style guide, she suggests encouraging talk about the project, giving a presentation, or sponsoring a contest. Washington also provides a two-page checklist for style guide development.

Another author who presents multiple steps to developing a style guide is Elizabeth R. Terlip. In "Writing standards: The foundations for new publications departments" (1989), she examines how corporate writing departments need to develop guides that create standards for document development, style, layout, and publication. She complains that "While most publication groups are struggling to keep pace with product development, it is difficult to find time to develop in-house documentation standards" (101). But she adds, "it is well worth the effort as a formal documentation standards guide reduces document organization time, minimizes style discrepancies, and better focuses your editing tasks" (101). This point reiterates similar arguments made by Caernarven-Smith, Washington, Allen, and others. Her steps for the development process include structuring standards, customizing standards, and implementing and updating standards.

Although Terlip refers to a "documentation standards guide" rather than a house style guide, the terms are basically synonymous. Among other things, as Terlip notes, standards can help structure the process of developing written or online documentation for corporate communication, include all or some of a corporation's documentation requirements, and ensure a consistent writing style and format. "A documentation standards guide," she contends, "is more than a style guide that explains when to use boldface or highlight headings" (101). Yet a style guide, by most definitions, also does more than those two functions.

Like Washington, she outlines what can be included in a guide. Her categories include:

- Document development procedures
- Page layouts and formatting instructions
- Style specifications
- File management and desktop publishing guidelines

These areas, she suggests, could constitute the chapters of a book. The section on style specifications is the longest, with thirty-three items under the subheading of text standards and six items under the subheading of graphics.

Other writers have offered their own schemes or variations on the considerations described above. Allen, who cites Washington for many points, envisions a five-step process:

1. Gaining management support
2. Defining the audience
3. Creating the style guide
4. Assembling the guide
5. Periodically updating the rules

Lalla suggests that procedures should depend largely on the circumstances, as discussed at various points in the preceding section.

Several questions occur, however. How empirically accurate are these steps when compared with what actually happens? Do some authors simply have a penchant for making lists? What support do the authors have for their claims? Why is there so much variance between different schemes? These questions have not been fully addressed in the literature to date.

House Style and Online Information

Another area that has not been fully explored is the relationship between house style and online information. Personal computers, online information, and new software programs have changed the field of written communications in many ways. The writing, editing, and publishing processes have been streamlined. Information can be sent quickly and efficiently over a modem line, and electronic journals are now becoming common. The Internet generally and the World Wide Web in particular have provided new opportunities in the technical communication field. Depending on the extent to which an organization is involved in producing online documents, the options, steps, and considerations for developing a style guide may need to be fine-tuned to account for this new and evolving medium.

In "Using style guidelines to create consistent online information," Michelle Corbin Nichols (1994) explains the need for rules of style for online documentation. Nichols confesses that she initially felt restricted by the style her company had adopted for creating online information, but eventually came to realize why such rules are needed. As with printed material, a consistent style for online documentation can help users easily locate the information they need. Effective online documentation, Nichols asserts, draws from the principles of many fields, including cognitive psychology, communications, and computer science.

According to Nichols, help screens and other online information should be easy to access, structured with different levels, specific to the situation, and concise. She suggests that information should proceed from general to detailed explanations, addressing the user's immediate questions and providing optional links to more detailed descriptions. Consistent screen designs are important so users can get data in the same way on different topics; in terms of writing, Nichols favors using similar syntactic structures (such as short, active sentences) throughout. Nichols says that making information easily scannable, formatting text in columns, and using lists and other devices can help make data more accessible to the reader. Thus, the notions of reader styles and audience are important, and style guidelines can help writers of online information appropriately frame their documents according to readers' expectations.

The relationship between computers and style issues is examined in a different way in "Corporate style" by Wayne Rash, Jr. (1991). Rather than looking at how technical writers must follow style rules, he examines how some software programs can help create a consistent style. At first glance, the article seems elementary: spell-check, grammar-check, and style-check programs are tools writers can use. A few vendors have even developed specialized editions, such as Grammatik IV Government Edition and Corporate Voice, to assist writers in specialized fields. Some programs can be further tailored to give prompts for situations unique to particular organizations. Because these programs supply general rules and can be further tailored for certain situations, they essentially provide a house style. The important point implied by Rash's article is that these programs are tools that can help the writer, in addition to the general style books already available. However, readers may not be sure whether Rash believes that such programs should ever completely replace a house style guide. I would suggest that a software program is only an aid, not a cure-all.

Voices of Dissent

Although many articles take a favorable stance toward establishing house style rules, there are occasional voices of dissent. The dissenters usually do not object to the concept of house style rules in general—these are not anarchists nor do they support an "anything goes" philosophy—but they protest what they see as arbitrariness or lack of flexibility in the enforcement of style rules.

Two articles, "Editors, authors, and audiences" by Douglas Wieringa (1995) and "Let the authors have their words" by Don Bush (1993), illustrate this perspective. Wieringa discusses the issues of decision-making in editing, grammar and style, and audience. He says that all editors have pet peeves about style issues, such as using the word *utilize* instead of the verb *use* or starting a sentence with a conjunction. The problem, however, is that many editors cross the line and end up distorting or completely changing the meaning of a passage. Wieringa argues that stylistic variations should be allowed and that jargon should be acceptable in some cases, such as using the word *grep* without any further definition for an audience of UNIX® users. Also, he notes that authors sometimes use literary devices to achieve certain effects, and these techniques may not always obey prescriptive grammar rules.

Bush addresses many of these same issues in his article, but he also raises a few more debatable points. He explores the issues of editing strategies, rigidity vs. flexibility, and suggestions for improving relations between authors and editors. Some of his points are convincing, such as trying to establish a good working relationship between writers and editors whenever possible. Overall, he demonstrates the need for professionalism and a consideration for authors' intent. He not only supports flexibility, but he suggests trying to encourage authors "to use their own departmental idiom" in communicating with a specific audience. He also supports deleting what he considers objectionable words like *approximately* before a phrase like *40,000 feet*, but allows a phrase such as *in situ*, regardless of whether it is clear to readers. To some, it may seem that he is being both rigid and flexible at the same time.

Flexibility is a tricky issue, and what many style guide opponents seem to forget is that it is a concept that varies from person to person. Thus, editors may believe they are being flexible, while others feel they are not. A style guide's purpose is to provide ground rules, with both the organization and the audience in mind. Although it is important to respect the author's purpose, one also needs to realize that there are many issues at work at the same time. If parts of a style guide need to be revised, many systems allow for that process. An effective, well-functioning system need not be perfect in all conceivable areas, but it should welcome change when change is necessary.

Conclusions

The current research makes clear that developing an effective house style guide requires much detailed planning. One must assess the organization's needs and sometimes juggle the expectations of management and staff. The matters of time allocation and funding should be taken into consideration. It is also necessary to recognize at the outset what a house style guide can and cannot accomplish, and to realize there are limitations to any printed document. Although the task of creating a style guide can easily become a long and daunting affair, the general consensus of the authors analyzed here seems to be that a carefully planned style guide is well worth the effort required to produce it.

This essay has presented a range of perspectives on issues relating to establishing a house style guide. Although some of the articles deal with specialized areas, they provide some insight into general areas of concern in style guide development. Although there is no single "right" way to create such a document, some themes appear repeatedly in these studies. There are also some areas of disagreement, such as whether to take a broad or comprehensive approach.

However, although many articles and conference presentations I have surveyed convey the authors' opinions on the effectiveness of style guides they have encountered, more concrete research data is needed on style guide development. Specifically, we need quantitative research that can be replicated and tested. Readers not only need to know *what works;* they would also like to know *when* something works. To sufficiently address such concerns would entail viewing matters more from the perspective of the ultimate users (rather than the designer) of a style guide. Too often, articles on style guide development have simply presented reflective narratives (on the part of the designer), without much consideration of whether the discussion of such experiences can be applied to other situations. Supporting data would enhance many of these studies. What is lacking is a basis for comparison. What is needed is a solid foundation of statistical evidence to support the claims made in these articles. For example, the actual results of a company's "needs analysis" study could be provided. Or statistics on style guide usage among practitioners could be determined through surveys and/or interviews. Studies could be done within a company, within a discipline, or within a geographical region—and, most importantly, results can be compiled and communicated to other interested professionals. Comprehensiveness is a concern as well, for many of the recent studies end almost as soon as they begin to shed some interesting light on an aspect of style guide development. The process of planning, creating, and implementing a house style is a topic that is largely uncharted and would prove ideal for further research.

References

Allen, Paul R. 1995. Save money with a corporate style guide. *Technical Communication* 42:284–289.

Blakely, J. Paul, and Anne S. Travis. 1987. The nine-year gestation of a unified technical style guide. *Proceedings of the 34th International Technical Communication Conference*, 63–64. Washington, DC: Society for Technical Communication.

Bush, Don. 1993 Let the authors have their words. *Technical Communication* 40:126–128.

Caernarven-Smith, Patricia. 1991. Aren't you glad you have a style guide? Don't you wish everybody did? *Technical Communication* 38:140–142.

Caruthers, Clifford M. 1986. The evolution of a style guide. *Proceedings of the 33rd International Technical Communication Conference*, 405–407. Washington, DC: Society for Technical Communication.

Chapman, Victor W., and Jean L. Owens. 1986. Finding solid ground: Using and articulating the grammar of technical editing. *Proceedings of the 32nd International Technical Communication Conference*, 67–68. Washington, DC: Society for Technical Communication.

Lalla, Sharon Trujillo. 1988. The state-of-the-art style guide development. *Proceedings of the 35th International Technical Communication Conference*, 176–179. Washington, DC: Society for Technical Communication.

Mitchell, Valerie. 1986. A style guide—Why and what. *Technical Communication* 33:232.

Nichols, Michelle Corbin. 1994. Using style guidelines to create consistent online information. *Technical Communication* 41:432–438.

Rash, Wayne Jr. 1991. Corporate style. *Byte* (February):89.

Terlip, Elizabeth A. 1989. Writing standards: The foundations for new publications departments. *Proceedings of the 36th International Technical Communication Conference*, 101–103. Washington, DC: Society for Technical Communication.

Washington, Durthy A. 1993. Creating the corporate style guide: Process and product. *Technical Communication* 40:505–510.

———. 1991. Developing a corporate style guide: Pitfalls and panaceas. *Technical Communication* 38:553–555.

Wieringa, Douglas. 1995. Editors, authors, and audiences. *Technical Communication* 42:101–103.

Commentary

In many respects, MacKay's article is a model literature review. Although his scope is modest, his treatment of the topic is comprehensive within that scope.

MacKay's research topic, corporate style guides, is an important and extremely useful one. Practitioners know that stylistic consistency is very difficult to maintain even for one person within a company, much less over the range of employees who write e-mail, memos, letters, procedures, user guides, online help, and the plethora of other types of written communication. Unfortunately, despite its importance, this topic has been largely neglected in the literature. For the ten-year period that MacKay has chosen to cover, only fourteen articles and conference papers addressed the subject, and most of those were very brief. In fact, of MacKay's fourteen references, only three were more than three pages long.

As a result, this short article in *Technical Communication* offers very detailed coverage of the literature on this subject, probably more so than would be possible for many topics.

Purpose and Audience

MacKay states his purpose succinctly in the final paragraph of his introduction: "This bibliographic essay presents an overview of recent research on style guides, critically analyzes these studies, and suggests areas for future consideration." This threefold purpose is common to all good literature reviews, and each part leads logically to the next. The writer first summarizes existing work on the topic, comments on its usefulness and value, and then proposes an agenda for future research on the topic.

Note that the "future research" aspect of a stand-alone literature review such as MacKay's is different from the corresponding focus of a review of literature found in an article, thesis, or dissertation. In that case, the conclusion would make the connection between the existing work on the subject and the author's own research project. The call for further research in such a work would be found in the concluding section of the article, thesis, or dissertation.

We can fairly assume that MacKay's audience is the audience of the journal in which his article appeared. *Technical Communication* is published specifically for the members of the Society for Technical Communication (STC) and more generally for others with an interest in the field. Since more than 90 percent of STC's members are technical communicators working in industry, we can assume that MacKay has directed his literature review to a practitioner audience. Indeed, the title of the article ("Establishing a corporate style guide") establishes its practical nature, and the works that MacKay addresses in his review are practitioner-oriented rather than theoretical. Furthermore, in the final paragraph of his introduction, MacKay expresses the hope that his article will be useful to "professionals who want to know what has been written on issues relating to style guide development."

Despite the practitioner focus, the article also implicitly appeals to academic readers in the conclusion's call for quantitative research (more typically undertaken by academics than by technical communicators in industry). In fact, the research that MacKay says remains to be done on corporate style guides offers great opportunities for projects that could team technical communicators in the workplace with academic colleagues.

Organization

The organization of MacKay's literature review is quite straightforward:

- Introduction (untitled)
- Planning Development: Determining Form and Content of a Style Guide
- Time, Financial, and Management Issues
- Steps for Developing a House Style Guide
- House Style and Online Information
- Voices of Dissent
- Conclusion

The introduction sets the stage for the rest of the article. It establishes the need for house style guides in corporate environments, provides a working definition of the terms *style guide* and *house style guide*, explains the organization of the essay, and describes the scope of the literature covered. MacKay's introduction offers a solid model for an article- or chapter-length literature review.

In the body of the article, MacKay examines each of the fourteen articles and papers he has selected to include in his literature review. His analysis is divided into five sections, each of which considers an aspect of corporate style guides, such as time, financial, and management issues, or the problems involved in providing style guidance for online information.

With two exceptions, he deals with each of the works in succession, first describing the content of the article or paper in a paragraph or two, and then sometimes providing an evaluation of the importance of the work or a critique of areas where it is lacking. At the end of each section he usually comments briefly on the points of commonality or difference among the works he has treated in that section.

In the section on "Steps for Developing a House Style Guide," MacKay refers to coverage of this aspect of the topic in the conference paper by Lalla (also discussed in "Planning Development: Determining Form and Content of a Style Guide") and in the articles by Caernarven-Smith and Allen (also discussed in "Time, Financial, and Management Issues"). These are the only sources mentioned in more than one section of the article's body. This fact is somewhat unusual as we would expect that articles and conference papers would typically address more than a single perspective on the topic. Because MacKay's sources are so brief, however, it is understandable that they do not take a more comprehensive approach to the larger topic of corporate style guides.

If MacKay's coverage of his sources can be faulted in any way, it is that he does not provide much in the way of evaluative comments about many of his sources. The fact that most of the papers and articles he covers are very brief and that he was a master's student at the time he wrote this article and was perhaps reluctant to criticize more established writers make this minor flaw understandable, however.

As noted earlier, the conclusion calls for quantitative research on style guides in corporate settings. Earlier in the article, MacKay took one of his sources, Paul Allen, to task for making statements about the money-saving potential of corporate style guides without providing any quantitative evidence to support that claim. MacKay's call for empirical research is well founded, but he should also include qualitative research in this plea. Much of the research on the value added by technical communicators (and that is essentially what we are talking about when we say that style manuals can save corporations money) has been qualitative rather than quantitative. And as we have seen in Chapter 5, qualitative research can make important contributions to the study of many aspects of our discipline.

Level of Detail

As we mentioned in the introduction to this commentary, authors are seldom able to present the rich level of detail that MacKay provides in his bibliographic essay on corporate style guides. Even in a stand-alone literature review such as this, the amount of previous research to be discussed will probably require that the author devote no more than a brief paragraph to each work included.

MacKay devotes four paragraphs each to his descriptions of three of his fourteen sources (Lalla, Allen, and Terlip); five others (Caruthers, Blakely and Travis, both articles by Washington, and Nichols) are covered in two paragraphs each. The remaining six works receive only a paragraph's attention each, though some of those paragraphs are quite lengthy. By contrast, the typical stand-alone literature review would probably devote no more than a paragraph each to discussions of most sources, and some might receive no more than a sentence or two. And the review of literature included in a primary research report is typically very brief. Most sources would be described in no more than a sentence, and less important sources or those that report similar results or make similar claims might be treated together in a single sentence.

Let us consider how the highly detailed description of one of MacKay's sources might be rewritten more concisely for inclusion in a more typical stand-alone bibliographic essay or a literature review that is part of a report on a qualitative research project on corporate style guides. Here are MacKay's two paragraphs describing Blakely and Travis's conference paper:

> In "The nine-year gestation of a unified technical style guide," J. Paul Blakely and Anne S. Travis (1987) outline the process and time required to develop a style guide for a scientific and engineering organization with 17,000 employees in four locations. Each plant had basic style guides, and manuals also existed for departments and divisions within departments at these plants; Blakely and Travis estimate that there were once 150 to 200 style guides in the company. The organization, Union Carbide, decided to consolidate its operations, and a committee was formed to develop a single, unified style guide. According to Blakely and Travis, because members of the committee had different professional backgrounds and interests, conflicts sometimes occurred, so a system of compromise and agreement had to be developed to keep the group focused. The authors note that the group's chairperson developed schedules and gave assignments to the members of the committee, and members had to report on assigned action items at weekly meetings. The group then debated various issues, arrived at decisions by voting, and presented proposals to upper management for approval, often a chapter at a time.
>
> The development process, Blakely and Travis explain, was a lengthy one. After several chapters were completed, drafts were distributed, and readers had the opportunity to provide their reactions on comment forms; the authors state that it took nearly 2 years to resolve all the comments and reach a final draft. Also, along the way, funding problems arose; the authors note that the project was nearly completed when funds ran out, and work stopped for about a year until a grant was received. From inception to completion, the entire development process lasted a total of 9 years, from 1977 to 1986. This presentation demonstrates that developing a style guide can be a monumental task.

By eliminating most of the details and examples, the essential information from MacKay's 300-word description can be reduced to 110 words:

> Blakely and Travis (1987) outline the process and time required to develop a style guide for a very large organization in four locations. When the company decided to consolidate its operations, a committee was formed to develop a single, unified style guide to replace up to 200 existing style guides. The chairperson developed schedules and gave assignments, and members reported at weekly meetings. The group debated issues, made decisions by voting, and presented proposals to management for approval. This part of the process took nearly 2 years, and when the project was nearly completed, funds ran out, and work stopped for about a year. The entire process lasted 9 years.

An even briefer version—only 57 words long—can be achieved by cutting the description to the bare facts:

> Blakely and Travis (1987) outline the process required to consolidate up to 200 existing style guides into a single, unified guide for the company's 4 locations. A committee met

weekly, debated issues, made decisions by voting, and presented proposals to management for approval. The entire process lasted 9 years, due in part to a loss of funding.

Finally, depending on the significance of the Blakely and Travis paper to the primary research being described in an article, the treatment might be ever terser (30 words):

> Chapman and Owens (1986), Blakely and Travis (1987), Caernarven-Smith (1991), and Allen (1995) discuss how time, finances, and other management issues can affect the decision to develop a style guide.

In this last example, all of the specifics about the Blakely and Travis paper have been removed, and it has been consolidated with three other sources on the same general topic.

Citations and References

Many academic writers tend to produce patchworks of quotations and details from the work of other authors, especially in literature reviews. Sometimes they do not seem to have understood what they have read, or if they have understood it, they do not convey that understanding to their readers. MacKay's audience certainly cannot make that charge about his literature review.

In this brief essay, MacKay uses only a few quotations from the articles and conference papers he addresses, and none of the quotations he includes is more than one sentence long. Instead of relying on quotation, MacKay uses paraphrase and (even more often) summary to convey to his readers what the literature has to say about the subject of corporate style guides.

The following paragraph illustrates the approach that MacKay takes to describing and analyzing his sources.

> In "A style guide—Why and what," Valerie Mitchell (1986) deals more with the content of style guides than with the dynamics of producing one. She says that a good style guide provides preferable word choices, determines what jargon is acceptable, sets rules to follow, and chooses authoritative dictionaries and reference sources. She, like Lalla, contends that a style guide should be comprehensive. Mitchell also suggests that it should deal with all aspects of developing publications. A style guide, she says, makes writers aware of their responsibilities and "ensures adherence to company standards, and at least *some* semblance of format" (232). Although Mitchell says that a style guide needs to promote only some semblance of format, other authors would hold that format should be a major concern. Mitchell concludes her brief article by providing a list of possible topics and subtopics for a style guide, including not only writing issues but also considerations of planning, scheduling, and formatting technical reports.

In the first sentence, MacKay provides an overview of the article, which deals with the content of a style guide. The remaining six sentences provide details supporting that overview, relying mostly on summary, with an occasional paraphrase and a single sentence of quotation from Mitchell's article.

1. Mitchell's article deals more with style guide content than with style guide production.
2. She explains that a style guide addresses word choices, jargon, rules, and authoritative reference sources.
3. Like Lalla, she believes that a style guide should be comprehensive.
4. It should deal with all aspects of the publication development process.
5. It should make writers aware of their responsibilities and ensure adherence to company standards, including minimum standards of format.
6. Some other authors would argue that format is more central to a style guide's purpose than Mitchell does.
7. Mitchell concludes with a list of topics a style guide should address.

Because he uses quotations and close paraphrases so sparingly, most of MacKay's citations are simply the article or paper's authors, title, and year of publication at the beginning of his discussion of that work. When he does quote from or closely paraphrase the literature, he ends the quotation or paraphrase with a parenthetical page reference using the Chicago style that is standard in *Technical Communication*. (Note that the citations and references in the version of the article reprinted here have been changed to conform to the 15th edition of the *Chicago Manual of Style*; *Technical Communication* used the 14th edition when the article was originally published.)

However, there are two instances where the page citations for quotations are missing. In his description of Clifford Caruthers's conference paper, MacKay has neglected to include a page reference for the quotation (405); similarly, he has omitted a page reference for the quotation from Don Bush's article (127). Although the author or the journal editor should have noticed these omissions before the article was published, they are minor problems.

Because MacKay has so thoroughly mastered the literature, his readers will find it very easy to achieve a similar high level of understanding of the works he discusses. Obviously, this is not a substitute for the detailed knowledge that an expert on the subject needs to attain, but it provides those who are already experts with a reminder of what they have previously read and gives novices an overview that will prepare them well for their own exploration of the sources he covers.

The list of references at the end of MacKay's literature review provides full bibliographical citations for each of the articles and papers he covers in the article. Had he referred to other works in the introduction, body, or conclusion of the article, those works would also be included in the list of references.

Exercise 7.1: Planning a literature review

Using the annotated bibliography on your research topic that you prepared in Exercise 3.2, construct an outline for the literature review for the paper or article you are planning. Your outline should indicate the major sections of your literature review; for each section, list each book and article that you would address in that section, followed by several bullets describing the major points you would want to include about that source. ■

Exercise 7.2: Planning coverage of a single work in a literature review

Suppose that you are Peter MacKay and are planning an updated version of your 1997 review of the literature on corporate style guides. Here is an entry from the annotated bibliography you have prepared on one of the sources that has appeared since your article was published.

> Bright, Mark R. 2005. Creating, implementing, and maintaining corporate style guides in an age of technology. *Technical Communication* 52:42-51.
>
> "This article details a step-by-step process for creating, implementing, and maintaining a corporate style guide to ensure consistency in organizational communication. Through literature research, analysis of sample style guides, and practitioner interviews, this article provides recommendations for gaining management support, building a process to develop a style guide, determining content, encouraging employee buy-in, and maintaining a corporate style guide." (http://www.ingentaconnect.com/search/article? title=style&title_type=tka&journal=technical+communication&journal_type=words&year_from=2001&year_to=2006&database=1&pageSize=20&index=6)
>
> Bright's article is significant because it expands the state of knowledge about developing corporate style guides based on comparative analysis of 17 examples, interviews with four practitioners who have developed style guides, and the author's personal experience. Although Bright's research is qualitative rather than quantitative, it provides a starting point for a more comprehensive study of the style guide as genre than any of the previous research, most of which has reported the experience of a single company or the opinions of a single style guide author or authoring team.

Would you include this article in one of the existing sections of the article or create a new section for it? Does the existence of Bright's article suggest other changes that might be needed? ■

Summary

This chapter has expanded on the concepts and methods presented in Chapter 3 by examining an article-length review of the literature about corporate style guides. Following the full text of the article, we have explored how its author went about writing it, examining its purpose and audience, organization, level of detail, and citations and references. Exercises at the end of the chapter provide practice in organizing a literature review and in planning how to incorporate a new source into a revision of the article we analyzed.

References

MacKay, P.D. 1997. Establishing a corporate style guide: A bibliographic essay. *Technical Communication* 44:244–251.

Answer Key

Exercise 7.1

Because the literature review outline produced for this exercise will be unique for each person who prepares it, there is no key for this exercise.

Exercise 7.2

Responses to this question will vary, but here are some possibilities.

A discussion of Bright's article might be added to the section on "Steps for Developing a House Style Guide," as the majority of the article deals with that aspect of corporate style guides. This section might be subdivided into two parts: one that is basically anecdotal (the section as it exists in the original article, plus any other anecdotal work on style guide development since 1997) and a second part that is more research-based (Bright's article and any other similar work on style guide development in the past ten years).

The discussion of Bright's article could also conceivably be added to the conclusion, with the conclusion changed to call for quantitative research and more qualitative research such as Bright's.

One major caution: Depending on the amount of material published on corporate style guides since 1997, significant structural changes may be necessary in the revised article. Revision may not simply be a matter of dropping new material into the existing outline.

8

Analyzing a Quantitative Research Report

Introduction

In Chapter 4, we explored quantitative studies in terms of their internal and external validity, as well as their reliability. We discussed hypothesis testing as a means of determining whether an intervention (also called an independent variable) makes a difference in terms of the results (or dependent variable) measured with two groups. We also considered descriptive statistics that provide information about a specific set of data, as well as inferential statistics that determine whether we can draw conclusions about a larger population based on the study sample. In this chapter, we will analyze a specific report about a quantitative research project to see how its authors have approached the task of exploring previous work on this topic. The chapter contains the full text of Michael Alley, Madeline Schreiber, Katrina Ramsdell, and John Muffo's "How the design of headlines in presentation slides affects audience retention," which appeared in the May 2006 issue of *Technical Communication*, as well as a detailed commentary.

Learning Objectives

After you have read this chapter, you should be able to

- Analyze a quantitative research report
- Apply the results of your analysis to reporting a quantitative study on your topic of interest

The Article's Context

As we noted in Chapter 4, hypothesis testing involving numerical averages is a common method in research because most people can easily relate to concepts and calculations involving averages. In the simplest type of hypothesis testing, the researcher formulates a hypothesis to test—for example, that exploded diagrams will help readers of a product assembly manual complete an assembly task more quickly than the same text with only illustrations of the assembled product to guide them. The researcher then tests the hypothesis by studying the performance of two groups of readers. One group receives the exploded diagram version (the independent variable), and the other receives the version with illustrations of the assembled product. The assembly time for each person (the dependent variable) is recorded, and the group to which the person belongs is noted.

The researcher analyzes the results of the study using descriptive and inferential statistics. The descriptive statistics used include the mean time each group took to assemble the product, as well as the standard deviation from the mean for each group. The researcher then knows whether there is a difference between the two groups in terms of assembly time. But the researcher also wants to know whether the difference is statistically significant. To make this determination, the researcher uses an inferential statistical tool that measures the probability that the difference is due to chance rather than to the intervention—the exploded diagrams.

The article we examine in this chapter is a very good example of hypothesis testing. Michael Alley and his colleagues want to determine whether the use of sentence headings in presentation slides (the independent variable) makes a difference in terms of the audience's retention of concepts presented. They designed a study to test that hypothesis, collected data, and analyzed it. The results of their study and the conclusions they drew from those results are reported in their article.

The article is important because it demonstrates a statistically significant difference in the retention of information (the dependent variable) by students taught with presentation slides that used sentence headlines as compared to the retention of students taught with slides that used word or phrase headlines. The findings are significant because they can likely be generalized beyond the academic environment to other learning situations in business and industry where presentation slides are commonly used.

How the Design of Headlines in Presentation Slides Affects Audience Retention*

Michael Alley, Madeline Schreiber, Katrina Ramsdell, and John Muffo

Introduction

The defaults for typography and layout in Microsoft PowerPoint, which has 95 percent of the market share for presentation slide software (Parker 2001), compel presenters to create headlines that are single words or short phrases. Not surprisingly, in a typical PowerPoint presentation, the main assertion of each slide is relegated to appear in the slide's body. For those creating slides for presentations, the question then arises: Is such a headline design the most effective at having the audience retain the slide's main assertion?

According to Robert Perry of Hughes Aircraft and Larry Gottlieb (2002) of Lawrence Livermore National Laboratory, the answer is "no." Since the 1960s, Perry has argued for a succinct sentence headline on presentation slides. Following Perry's lead, Gottlieb came to the same conclusion during the 1970s at Lawrence Livermore Laboratory. For the next three decades, although a number of technical communicators strongly advocated using sentence-headline designs, the overwhelming majority of headline designs projected at engineering and scientific conferences were single

* This article was originally published in 2006 in *Technical Communication* 53:225–234. Reprinted with the permission of the authors.

words or short phrases. Recently, in the midst of complaints from popular media (Parker 2001; Schwartz 2003) about the use of PowerPoint in presentations, several publications, including Alley (2003a), Doumont (2005), and Atkinson (2005), repeated the old arguments and presented new ones for using headlines that are sentences.

Sentence headlines have several main advantages over phrase headlines (Alley and Neeley 2005). First, a sentence headline such as *Placer deposits arise from the erosion of lode deposits* orients the audience much more effectively to the slide's purpose than does a phrase headline such as *Placer Deposits*. Second, using sentence headlines allows the presenter to emphasize the most important detail of the slide. Third, if well chosen, sentence headlines present the audience with the key assertions and assumptions of the presentation. Explicitly stating these assertions and assumptions in a technical presentation is advantageous because audiences are more inclined to believe the presentation's argument if they comprehend the assertions and assumptions of that argument (Toulmin 2003). Finally, once the headline assertion has been determined, the presenter is in a much better position to select persuasive evidence to support that assertion.

This article presents an experimental study on the effect of sentence headlines in four sections of a large geoscience course that typically had 200 students per section. In the study, the four different sections of students were taught the same information by the same instructor, with the only difference being the design of the teaching slides.

Of the four sections of students, two sections viewed the information on slides that used mostly phrase headlines (note that some of these original headlines were formatted as questions, and a few slides did not have any headlines). The remaining two sections viewed the same information on slides that used succinct sentence headlines. In the slide transformations, other changes occurred, such as typographical changes and conversions of bullet lists to more visual evidence. However, for the 15 slide transformations considered in this study, the principal change was the conversion of a traditional headline to a succinct sentence headline.

After each class period, all four sections of students had access to copies of the slides that the instructor had projected. Then after the five class periods, the students took an exam that asked them to recall a set of assertions from those slides. For those in the two sections taught from the traditionally designed slides, the assertions resided in bodies of the slides, while for the students in the sentence-headline sections, the same assertions resided in the slides' sentence headlines. The course's final examination, which occurred a few days after the final class period, served as the recall test.

This case of an audience viewing a set of slides and then having access to those slides as a set of notes is common in science and engineering. Granted, the way that students study their set of notes for a final exam is quite different from the way that technical professionals would refer to their sets of notes. Nonetheless, the results presented here have implications in the way that technical professionals should design slides.

For instance, if the students who were taught from the slides with sentence headlines recalled significantly more information than those students who were taught from slides with phrases, questions, or no headlines, then technical presenters should consider using sentence-headline designs. In such a case, given that the overwhelming majority of technical presenters currently use phrase headlines, the increase in the amount of technical information communicated in engineering and science could be large.

The next section of this article describes the design of the study. Included in this section is a justification of the particular sentence-headline design selected for the study, the control method used to assess the relative strengths of the four student groups, and a key assumption about the tests. Following this section are the study's results. At the heart of this section is an explanation, from a communication perspective, of why the students who viewed the sentence headlines recalled the slides' key assertions at levels that were different from those who viewed slides with phrase, question, or no headlines.

Experimental Methods

This study considered the effect on audience retention of using a sentence-headline design for the teaching slides in a large geoscience course at Virginia Tech. This was an introductory course that discussed the origin, distribution, and use of the Earth's resources. Because the course satisfied one of the university's general education requirements, it was a popular course for non-majors, attracting students from all branches of science and engineering, as well as those from liberal arts, agriculture, and business.

The course was excellent for this pilot study because the instructor used computer-generated projections of slides as the principal visual aid in most class periods. For that reason, the slides played an important role in the instruction. Other reasons that the course was a good choice for this study were that the examinations had multiple-choice questions, the students took examinations on sheets that could be computer scored, and the instructor had examinations graded through the university testing center, where the results of prior examinations exist. From these examination results, we were able to extract statistics directly linking test questions to presentation slides from earlier semesters. Shown in Figure 8.1 is a visual depiction of how the study was performed.

For the study, we transformed about 100 teaching slides from the fourth and final portion of the course to the sentence-headline design. Not all the transformations involved the same types of changes. In the instructor's original design of slides, about 80 percent of the slides had phrase headlines; the remaining 20 percent either had no headlines or had headlines written as questions. In the transformed versions, 100 percent of the slides, except for the title slide of each class period, contained succinct sentence headlines.

In addition, about 40 percent of the original slides consisted of the traditional bullet list in the body, with the remaining 60 percent having at least one image. In the transformed versions, 100 percent had the evidence of its slide bodies presented in a visual way without any bullet lists being used. No doubt these visual changes to the slide bodies affected audience recall (Alley, Schreiber, and Muffo 2005). However, for the 15 slide transformations tested for this study, the principal change was the conversion to a succinct sentence headline. Moreover, for those 15 transformed slides, the assertions that the students had to recall resided in the sentence headlines, while for the corresponding traditional slides, those assertions resided in the bodies of the slides.

Using the transformed slides, the instructor taught the classes in the same way that she had done in past semesters. Of particular importance, as she had done in

Figure 8.1 Visual depiction of the strategy for the study. The isolated difference between the presentations of the information was the design of the slides.

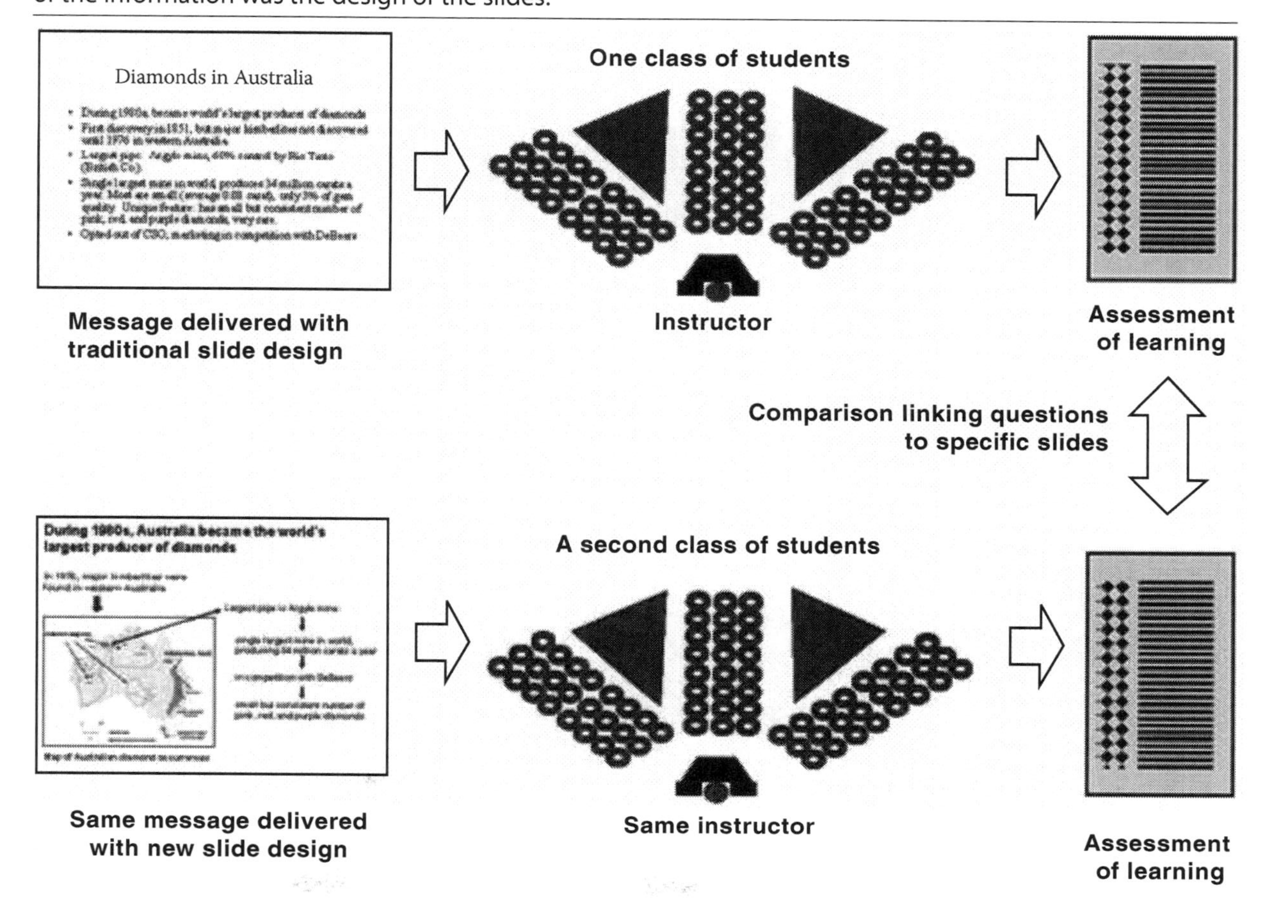

past semesters and in the other lectures that semester, she posted her slides on the Web so that students could download the slides after the lectures to use as study aids.

Justification of selected design. The sentence-headline design chosen for this study has achieved much anecdotal success (Alley and Neeley 2005). One feature of this design is its set of specific guidelines for typography (Alley 2003a). For instance, one such guideline is that the sentence headline be restricted to no more than two lines. This guideline agrees with Doumont's recommendation (2005) for text blocks on a presentation slide. A second typography guideline is the use of a bold sans-serif typeface for the headline. This guideline arose from our own observation that a boldface sans-serif typeface is easier to read, in a large room, than either a normal sans-serif or a normal or boldface serif formatted *at the same type size.* In assessing the ease of reading for different typefaces in the room, we positioned ourselves both at the back-row seats most distant from the screen and at the front-row seats with the sharpest angles to the screen. Yet a third typography guideline was left justifying the headline with a beginning position in the slide's upper left corner. This guideline agrees with the recommendations of Gottlieb (2002) for sentence headlines.

A second feature of the slide design chosen for this study is that the headline be supported by visual evidence, as opposed to a bullet list. This aspect agrees with one of Richard Mayer's principles (2001) for multimedia—namely, that students learn better from words and representative images than from words alone. The slide design chosen for our study also follows two more of Mayer's principles: (1) that students learn better when images are placed near rather than far from the corresponding text; and (2) that students learn better when images and corresponding text are presented simultaneously rather than successively. The slides of Figure 8.2 show the differences between the traditional design (top) and the sentence-headline design selected for this study (bottom).

Control group for the study. The final examinations for the four different sections of the course consisted of 100 questions: 60 questions based on the content for the course's fourth and final portion, and 40 questions drawn from the questions already posed to the students on the semester's three earlier tests. Given this structure, we chose the average score that each class had on the 40 questions from the previous tests as a means for controlling the relative effort of each section. We chose the average scores for these 40 questions as the control measure because students in all four sections prepared for this portion of the examination by studying the previous tests of the semester rather than by studying the slides that accompanied that material. For that reason, the results on this section of the exam provided an excellent window into the effort given by each section of students.

Table 8.1 presents a summary of the averages obtained by each of the four sections for those 40 questions. As seen in Table 8.1, the Fall 2004 and Spring 2005 sections achieved lower scores than the Fall 2003 and Spring 2003 sections did. For that reason, we concluded that the two later sections, which were taught with sentence-headline slides in the fourth and final portion of the semester, did not put forth as much effort in preparing for the exam as did the two earlier sections, which were taught the same material from slides designed in a traditional way.

Figure 8.2 Transformation of one of the traditionally designed slides, shown at the top, to the sentence-headline design shown at the bottom (Schreiber 2005).

Diamonds in Australia

- During 1980s, became world's largest producer of diamonds
- First discovery in 1851, but major kimberlites not discovered until 1976 in western Australia
- Largest pipe: Argyle mine, 60% control by Rio Tinto (British Company)
- Single largest mine in world, produces 34 million carats a year. Most are small (average 0.08 carat), only 5% of gem quality. Unique feature: has small but consistent number of pink, red, and purple diamonds, very rare
- Opted out of CSO, marketing in competition with DeBeers

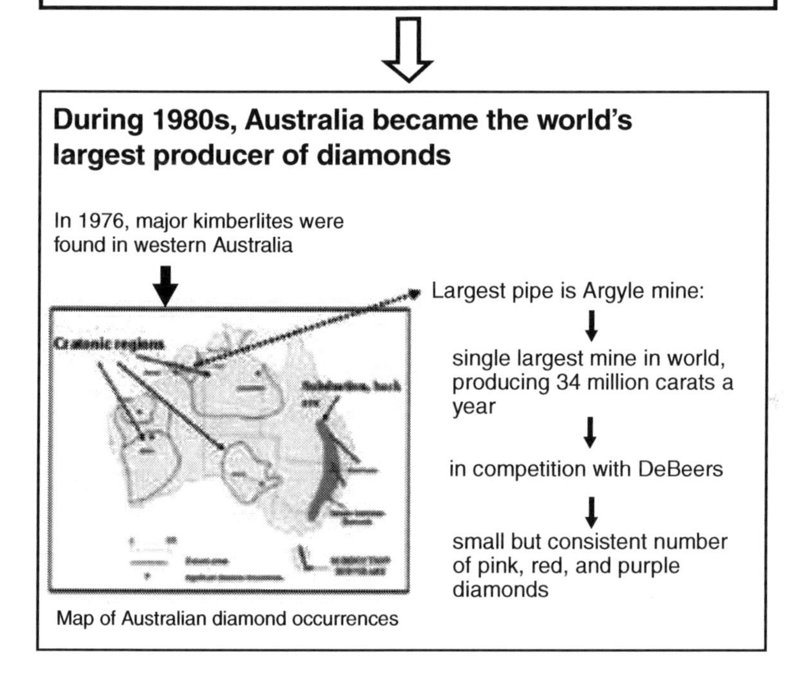

Key assumption in the study. Richard Mayer (2001) has performed several experimental studies on the effect of multimedia on learning—learning from words and pictures as opposed to learning from just words. In Mayer's studies, each learner received exactly the same words (either written, spoken, or both) because the spoken words were narrated, rather than presented. Mayer's studies provide a valuable base of knowledge on how words (written and spoken) and images affect

Table 8.1 Scores for Different Sections on Control Questions

Section's Semester	No. of Students	Class Time	Class Score: Control Questions
Spring 2003	200	2:30–3:45 PM	87.9%
Fall 2003	202	12:30–1:45 PM	86.1%
Fall 2004	201	12:30–1:45 PM	82.5%
Spring 2005	136	8:00–9:15 AM	79.1%
Average			83.9%

how much people understand and recall. However, the communication situations of Mayer's studies are quite different from the communication situations that most technical professionals face.

For instance, because the speech in Mayer's studies was recorded and played to the audience, the speech was perfect—exactly what the presenter wanted the audience to hear. In contrast, in a typical technical presentation such as the presentation of scientific research at a conference or of an engineering design to managers, the presenter speaks from points or slides. Because not every word is scripted, the wording is not exactly the same from one presentation of the material to another. Moreover, the speech is sometimes influenced by the audience—either the presenter reacting to the expressions of the audience or, during an informal presentation, the speaker responding to questions interjected by the audience. In addition, in most technical presentations, the presenter not only has to inform the audience about the information but also has to persuade the audience about that information. For that reason, the presenter has to build credibility with the audience. In building that credibility, the presenter often includes details that are beyond what the audience needs to understand the content—a strategy that goes against Mayer's principle of excluding extraneous words (2001).

The study presented in this article is much more like the typical situation that technical presenters face in that the presenter delivered her class live, rather than taped. Given that live element, though, each section did not experience exactly the same speech. Moreover, in our study, students in different sections asked different questions, which caused some points to be emphasized more than others. Finally, full attendance did not occur during every class, which meant that some students learned portions of the material just from the posted slides, as opposed to learning the material from both the classroom presentation and the posted slides.

Despite these irregularities in the speech experienced by the audience, a key assumption of this article is that the design of slides was the major difference in the learning that occurred among the four sections: the Spring 2003 and Fall 2003 sections that learned from slides mostly with phrase headlines, and the Fall 2004 and Spring 2005 sections that learned from sentence-headline slides. Put another way, the students in these large sections experienced, on average, the same speech. Supporting this assertion is the overall consistency in test scores in which the different sections witnessed the same information presented on the slides in essentially the same way and answered the same questions that arose from those slides.

Examples of consistency include the six exam questions that required the students to recall information *from images on slides* that could be found on both the traditional slides and the sentence-headline slides. For these questions, the scores were close, with an average correct score of 86% for those students learning from the traditional slides and 87% for those students learning from the transformed slides—a difference that is not statistically significant. That the scores were so close is not surprising because the transformations did not make as much of a difference on these slides. The information to be recalled was not directly stated in the sentence headline, and the images were the same.

Other examples of consistency include seventeen exam questions that required the students to recall information from text that was in the bodies of both the traditional slides and the sentence-headline slides. In the transformations corresponding to these seventeen questions, although sentence headlines were added to the transformed slides, those headlines did not contain the information to be recalled. In addition, these transformations did not involve significant reworking of text into visual arrangements—key images already existed on the slides. Moreover, the amount of text on the transformed slides was about the same as on the original slides. For these questions, the scores were also close, with an average correct score of 73% for the students learning from the traditional slides and 74% for the students learning from the transformed slides—a difference that is not statistically significant.

What these two sets of data reveal is that when students from the different sections were asked to recall information that was in the slide's body and incorporated about the same way—as an image or as body text—the test scores were about the same.

Results and Discussion

Fifteen questions from the final exams required the students in either the Fall 2004 or Spring 2005 sections to recall information that existed in one of the slide's sentence headlines. For the Fall 2003 or Spring 2003 sections, the questions required students to recall the same information—the difference for these two earlier sections was that information existed within the text of a slide's body. The average score for the students taught from the traditional slides was 69% correct, while the average for the students taught from the slides with the sentence headlines was 79% correct. A chi-square analysis shows that this difference is statistically significant at the 0.001 significance level.

On seven of the fifteen questions, the students viewing the sentence-headline slides achieved higher scores that were statistically significant (three at the 0.001 significance level, three at the 0.005 significance level, and one at the 0.025 significance level), as opposed to achieving lower test scores that were statistically significant on only two questions (both at the 0.01 significance level). Larger sample sizes might have yielded significant differences on several other questions.

Table 8.2 presents a comparison for those fifteen questions of the test scores between a section that was taught from slides with the traditional headlines and a section that was taught from slides with sentence headlines. Note that three questions appear twice in the table: questions 3 and 5 are the same, questions 6 and 7 are the same, and questions 8 and 15 are the same. These questions were posed either to two different sections that viewed the sentence-headline slides or to two different sections that viewed the slides with the traditional headlines.

Figure 8.3 presents a graph of these same statistics. As shown, for seven of the fifteen questions, the group learning from the sentence-headline slides achieved test

Table 8.2 Comparison of Test Scores for Those Taught from Traditional Headlines Vs. Scores for Those Taught from Sentence Headlines

Question	Original Form of Headline	Percentage Correct for Traditional Headline	Percentage Correct for Sentence Headline	Significance Level of Statistical Difference
1	None	23	57	0.001
2	Question	24	58	0.001
3	Phrase	61	85	0.001
4	Phrase	46	63	0.005
5	Phrase	71	85	0.005
6	Phrase	75	89	0.005
7	Phrase	79	89	0.025
8	Phrase	79	86	not significant
9	Phrase	80	85	not significant
10	Phrase	74	79	not significant
11	None	67	72	not significant
12	Question	96	99	not significant
13	Phrase	86	81	not significant
14	Question	96	89	0.01
15	Phrase	79	63	0.01
Average		69	79	0.001

scores that were significantly higher than the scores achieved by the group viewing the slides with phrase, question, or no headlines.

For ten of the fifteen questions, the transformation involved changing a phrase headline to a sentence headline. Such transformations correspond to Questions 3-10, 13, and 15 in the data. Shown in Figure 8.4 is a comparison of test scores for one such transformation. In this case, the students taught from the traditional slide were asked to recall the information given in the first bullet point, while the students taught from the transformed slide were asked to recall the information given in the sentence headline. The test score for the student group taught from the phrase-headline slide was 46%, while the raw score for the student group taught from the sentence-headline slide was 63%—a statistical difference at the 0.005 significance level.

What led to such a difference? Certainly contributing to the increased recall was the greater typographical emphasis given to the information in the sentence headline as opposed to the information provided by the body text of the slide with the traditional design. The larger type size (28 points vs. 24 points), the use of boldface, the placement of the detail at the top of the slide—all of these placed more emphasis on the detail in the sentence headline. In addition, on the traditional slide, the placement of the detail in a bullet list reduced emphasis on that detail, even though the bullet point was the first one listed. As Shaw, Brown, and Bromiley (1998) assert, bullets serve to remove hierarchy given to details. For that reason, this detail in the

Figure 8.3 Ratio of the raw test scores for the group that was taught from the sentence-headline slides over the raw test scores of the group that were taught from slides with traditional headlines. Light gray bars on the left represent significant increases, black bars represent differences that were not significant, and the dark gray bars on the right represent significant decreases (significance levels given above) the bars.

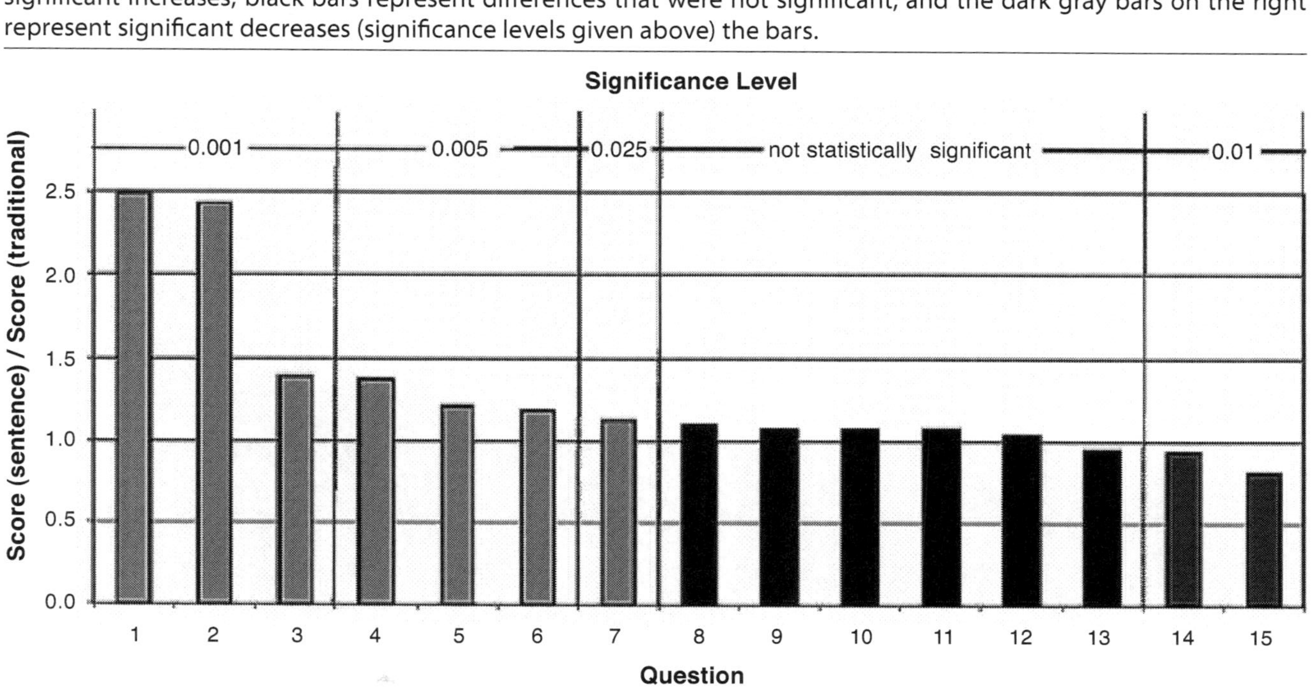

Figure 8.4 Comparison of test score of 46% correct for a slide with a phrase headline, shown at the top, with a test score of 63% correct for the sentence-headline slide on the bottom (Schreiber 2005). The test question asked the students to recall how much iron is in the Earth's crust.

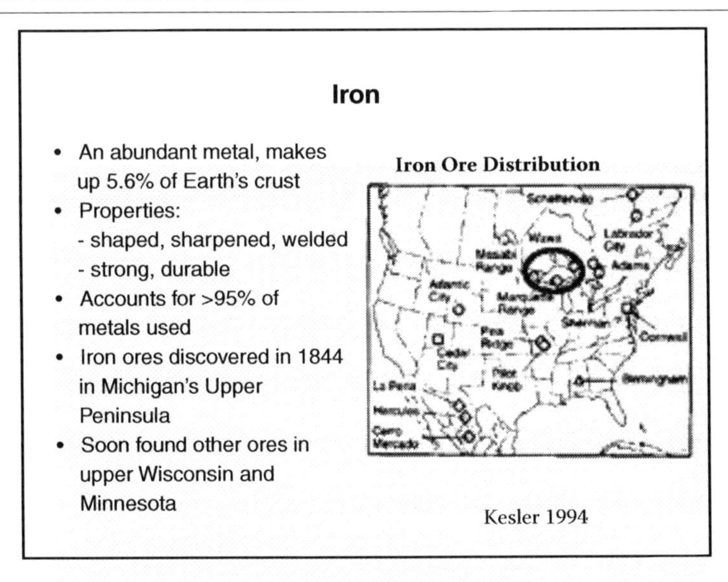

Led to 46% recall

⇩

Led to 63% recall

list was emphasized about the same as the list's less important details, such as the discovery date of ores in Michigan's Upper Peninsula.

Yet a third reason for the increased recall in the transformed slide was subordination of two less important details to the presenter's speech. Although the students learning from the bottom slide heard all the details from the original slide in the presentation, those students were not burdened with those less important details when viewing the slide either during or after the presentation. This design choice of having the subordinate information removed from the slide follows Mayer's (2001) principle that students learn better when extraneous words are excluded.

Also perhaps contributing to the increased recall for the students taught from the sentence-headline slide was the way in which the students cataloged the detail in their memories. Because the students' orientation to the slide was through an assertion that contained the detail, that is perhaps how they cataloged the information—anchored to that detail. In other words, the detail was cataloged as a first-level detail. That sort of cataloging would contrast with the way that the students viewing the phrase-headline slides might have cataloged the information. The phrase-headline students might have cataloged the detail as a second-level detail beneath the first-level heading of *iron*. As a second-level detail, it was less likely to be recalled.

For five of the ten questions that involved transformations of phrase headlines to sentence headlines, the students taught from the sentence-headline slides achieved higher scores that were statistically significant. In turn, on only one question did these students achieve lower scores that were statistically significant. The one question for which there was a significant decrease in recall occurred with the Spring 2003 section (phrase headline) achieving a significantly higher score than the Fall 2004 section did (sentence headline). Interestingly, on that same question, the Spring 2005 section (sentence headline) actually achieved a higher score than did the Spring 2003 section (phrase headline). In effect, although the Fall 2004 and Spring 2005 students were taught from the same slide on this question, the Spring 2005 students scored much higher than the Fall 2004 ones did. As mentioned, when students from different sections viewed the same slides and were asked to recall the same information from those slides, the scores were generally about the same. However, this case was clearly an exception.

For two of the fifteen questions, the transformation involved changing a slide with no headline to a sentence headline. Such transformations correspond to Questions 1 and 11 in the data. Figure 8.5 shows the difference in recall that occurred in the transformation corresponding to Question 1. In this case, the students taught from the slide with no headline were asked to recall the information given in the body of the slide, while the students taught from the transformed slide were asked to recall the same information given in the sentence headline. The test score for the student group taught from the slide with no headline was 23%, while the raw score for the student group presented with the sentence-headline slide was 57%—a statistical difference at the 0.001 significance level.

What led to the difference in recall in this slide was certainly the increased typographical emphasis given to the detail in the sentence headline. Perhaps also contributing was that the students taught from the sentence-headline slide cataloged the detail in the sentence-headline slide as a first-level detail, while the students viewing the slide without a headline did not have a memory anchor for this detail.

Figure 8.5 Comparison of test score of 23% correct for a slide with no headline, shown at the top, with a test score of 57% correct for the sentence-headline slide on the bottom (Schreiber 2005). The test question asked the students to recall where the Crandon ore formed.

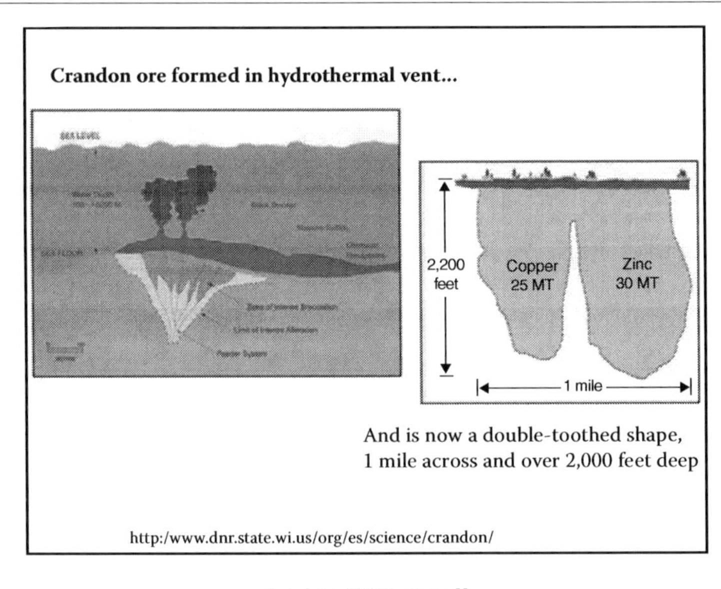

Led to 23% recall

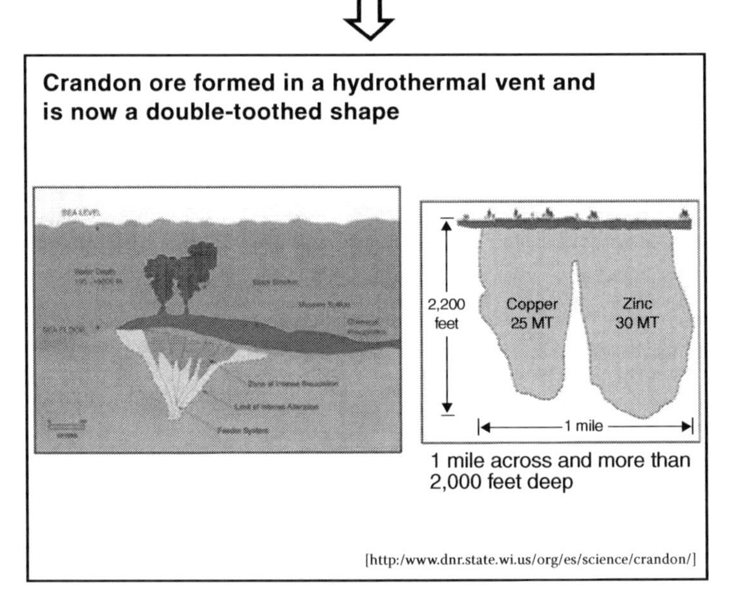

Led to 57% recall

For three of the fifteen questions, the transformation involved changing a slide with a question headline to a sentence headline. Such transformations correspond to Questions 2, 12, and 14 in the data. Shown in Figure 8.6 is the transformation for the slide corresponding to Question 2. In this case, the students taught from the slide with the question headline were asked to recall the information given in the text, while the students taught from the transformed slide were asked to recall the information given in the sentence headline. The raw score for the section learning from the question-headline slide was 23%, while the raw score for the section learning from the sentence-headline slide was 57%—a statistical difference at the 0.001 significance level.

In comparison with arguments for using a phrase headline or having no headline, the argument for using a question headline is stronger: a question headline leads the presenter to introduce the topic in an active way. In other words, if the presenter poses the question headline to the audience, the audience is challenged to seek the main assertion of the slide. However, on Question 2, the students taught from the question-headline slide performed much worse than did the students taught from the sentence-headline slide. One likely reason was that much synthesis was required of the students who viewed this particular question headline. In effect, these students had to recall three details from the body, and one of those details (the detail about irradiation) was not grouped with the other two (impurities and defects). For the students viewing the slide with the sentence headline, though, those three elements were grouped into one assertion.

On the surface, the original slide appears to be weak, since the positions of the main assertion's details are fragmented in the slide's body. However, such fragmentation is not uncommon for slides that rely on traditional headlines. It is not until the presenter identifies the main assertion—a step that creating sentence headlines ensures—that the presenter can clearly see what those details are and whether those details are arranged effectively.

Although this particular question headline was much less effective than the sentence headline, situations arise in which a technical presenter might consider using a question headline in series with a sentence headline. One such situation would be when the presentation would benefit from the audience examining the evidence in the body of the slide before seeing the assertion, as in the presentation of an assertion for which the audience has a hostile reaction. In such a use, the question headline would appear first, and then after the presenter has addressed the question by examining evidence in the slide's body, the presenter would "animate in" the sentence headline.

Conclusions

According to Microsoft (Parker 2001), an estimated 30 million PowerPoint presentations occur every day. As anyone who has recently attended a conference knows, the overwhelming majority of those presentations have headlines that are either single words or short phrases. This article, though, has presented experimental evidence that succinct sentence headlines are more effective. In our study, using sentence headlines of no more than two lines led to statistically significant increases in recall from the audience on details that were contained in those sentence headlines. A chi-square analysis shows that this difference is statistically significant at the 0.001 significance level. The main conclusion is that if technical presenters desire to

Figure 8.6 Comparison of test score of 24% correct for a slide with the question headline, shown at the top, with a test score of 58% correct for the sentence-headline slide on the bottom (Schreiber 2005). The test question asked the students to recall what caused color in diamonds.

What causes color in diamonds?

Fancy color is rare

Colors come from impurities or defects

Examples:

Yellow: nitrogen

Blue: boron

Green: uranium (irradiation)

Red/Pink: unknown

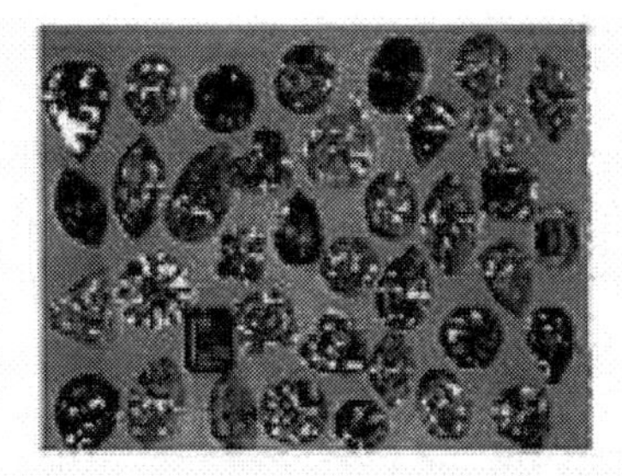

Led to 24% recall

In diamonds, colors (which are rare) come from impurities, defects, and irradiation

Yellow: nitrogen

Blue: boron

Green: uranium (irradiation)

Red/Pink: unknown

Aurora Collection of naturally colored diamonds [http://www.amnh.org]

J. Lo's ring from B. Affleck

Led to 58% recall

emphasize assertions in a presentation, they would do well to place those key assertions in succinct sentence headlines.

For a presenter desiring to design slides with such sentence headlines, however, the typography and layout defaults of PowerPoint pose a hurdle. In other words, someone who simply opens up PowerPoint must make many keystrokes to change the size, position, and alignment of the headline's text to accommodate a sentence appropriate for a technical presentation. However, help does exist. For instance, available at the first Google listing for the topic of *presentation slides* are templates to overcome the cumbersome headline defaults of PowerPoint (Alley 2003b). For many technical presenters, these templates have made the adoption of a sentence-headline design much easier (Alley and Neeley 2005).

This pilot study has focused on slide transformations in which the principal change on each slide was to place the slide's main assertion into a succinct sentence headline. The research question for this study was how well the audience has retained that assertion. Further testing is needed to isolate completely the effect of changing the sentence headline from other aspects modified in the slide's design—in particular, the slide's typography or the way in which evidence in the slide's body was designed.

Acknowledgments

Our thanks go to the Center for Excellence in Undergraduate Teaching at Virginia Tech for their financial support and advice. In particular, we wish to acknowledge Professor Terry Wildman and Meghan Habas Siudzinski.

References

Alley, Michael. 2003a. *The Craft of Scientific Presentations.* New York, NY: Springer-Verlag, 125–129.

———. 2003b. Design of presentation slides, http://writing.eng.vt.edu/slides.html.

Alley, Michael, and Kathryn A. Neeley. 2005. Rethinking the design of presentation slides: a case for sentence headlines and visual evidence. *Technical Communication,* 52:417–426.

Alley, Michael, Madeline Schreiber, Katrina Ramsdell, and John Muffo. 2005. Pilot testing of a new design for presentation slides to teach science and engineering. *Proceedings for the 35th ASEE/IEEE Frontiers in Education Conference.* Indianapolis, IN: IEEE.

Atkinson, Cliff. 2005. *Beyond Bullet Points: Using Microsoft PowerPoint to Create Presentations That Inform, Motivate, and Inspire.* Redmond, WA: Microsoft Press.

Doumont, Jean-luc. 2005. Slides are not all evil. *Technical Communication,* 52:64–70.

Gottlieb, Larry. 2002. Well organized ideas fight audience confusion. *49th Annual Conference of the Society of Technical Communication.* Arlington, VA: Society for Technical Communication.

Mayer, Richard E. 2001. *Multimedia Learning.* New York, NY: Cambridge University Press.

Parker, I. 2001. Absolute PowerPoint. *The New Yorker* (28 May):76–87.

Schreiber, Madeline. 2005. Teaching slides from Resources Geology: Geos 1024. Blacksburg, VA: Department of Geosciences, Virginia Tech.

Schwartz, John. 2003. The level of discourse continues to slide. *The New York Times,* 28 September, sec. 4, 12, col. 1.

Shaw, Gordon, Robert Brown, and Philip Bromiley. 1998. Strategic stories: How 3M is rewriting business planning. *Harvard Business Review* (May–June):41–50.

Toulmin, Stephen E. 2003. *The Uses of Argument.* New York, NY: Cambridge University Press.

Commentary

We selected this study by Alley and colleagues for three primary reasons. First, it is a brief, straightforward example of an article reporting the results of hypothesis testing. There are no significant problems with the design or execution of the study, nor with the analysis or reporting of the results.

Second, although the article reports the results of hypothesis testing, it does not use the same statistical test we illustrate in Chapter 4. For the most part, it does not compare means or averages, and it uses the chi-square statistical tool rather than the *t*-test. We hope that by seeing an article with a different statistical test, you will see that the principles of critically evaluating a quantitative study remain the same. As we stated in Chapter 4, the *t*-test is simply one statistical instrument in the researcher's toolkit, and it is used to compare a mean to an expected value. Because Alley and colleagues are comparing the percentage of correct answers between two groups rather than average scores, they could not use the *t*-test. Instead, they use the chi-square, which measures whether the differences in percentage of correct answers between the two groups varies by more than the amount that would be expected by coincidence. This approach allows them to make a much finer analysis of the differences between the two groups than the simple comparison of the average scores of the two populations on the questions of interest.

Finally, we chose this article because it deals with a topic that is relevant to virtually all technical communicators—the relative effectiveness of two types of presentation slide headlines. Although the study itself deals with slides used in the context of a university course, the results can legitimately be generalized to virtually any environment where slides are used to support the spoken presentation of information by a speaker and later to provide a reminder of that information to those who heard the presentation.

Purpose and Audience

The purpose of this article is to explore the authors' research question, stated in the last sentence of the first paragraph: "Is … a headline design [consisting of single words or short phrases] the most effective at having the audience retain the slide's main assertion?" The purpose is restated in greater detail in the article's fourth paragraph:

> This article presents an experimental study on the effect of sentence headlines in four sections of a large geoscience course that typically had 200 students per section. In the study, the four different sections of students were taught the same information by the same instructor, with the only difference being the design of the teaching slides.

Because the article was published in *Technical Communication*, we can safely assume that its primary audience is technical communication practitioners, with a secondary audience of technical communication teachers. Although the study was conducted with a population of students enrolled in a university course and the researchers themselves are three professors and a student, the authors convincingly demonstrate that their results have value beyond the academic environment. Indeed, the significance of the article lies in the generalizablity of their results.

Organization

The organization of this article is a variation on the standard IMRAD format for research reports: Introduction, Methods, Results, and Discussion.

- In the Introduction section, Alley and his colleagues provide the general context for the study (note the literature review in the first three paragraphs), state their research question, and describe when and where they conducted the study.
- In the Experimental Methods section, the authors provide the details of their experimental design. In this section, we learn how the slides presented to the control group were transformed for presentation to the experimental group. The researchers also describe the examination used to test audience retention of the information presented on the slides.
- The authors combine the final two sections of the IMRAD structure in their Results and Discussion section. Here, they provide details about the responses to the fifteen examination questions whose answers were provided in the body of slides presented to the control group and in the sentence headlines of slides presented to the experimental group.
- The Conclusion restates the overall findings, generalizes them to a wider audience, and provides further insights.

The IMRAD format is commonly used by authors of quantitative research papers and articles in medicine, the physical sciences, and engineering, as well as in the social sciences. The fact that Alley and his coauthors use this format is not surprising. In fact, their adoption of the IMRAD structure might even be said to give the article additional credibility because it observes the conventions for presentation of quantitative research.

Study Design

You will recall that one of our examples in Chapter 4 proposed to examine how incorporating the results of usability testing in a Web site reduced the mean registration time between users of the original design and users of the revised design. We formulated a null hypothesis ("There will *not* be a statistically significant reduction in the mean registration time ..."), collected data with users of both versions of the site, calculated the mean for both sets of users, and tested the difference between the means for statistical significance. Alley and his colleagues took a very similar approach in the research that they report in their article, but the differences between our simple example in Chapter 4 and the "real-world" example reported in the article we are examining here are instructive.

Hypothesis. In Chapter 4, we explained that you should begin a research project by formulating a hypothesis based on your research question that states the independent variable (the intervention you will test), the dependent variable (the result you will measure), and the predicted direction of the change you expect to see ("a statistically significant reduction in registration time," for example). We also noted that the hypothesis is usually stated in the null form: "There will be *no* statistically significant reduction in registration time between users of the original Web site and users of the

Web site redesigned using the results of usability testing." This hypothesis is usually stated explicitly in the research report.

Although it is important for researchers to understand that statistical reliability comes from testing the null hypothesis, many published reports do not explicitly state the test and null hypotheses. One reason is that published reports must speak somewhat plainly to a more general audience. That is the case with the article by Alley and his colleagues: The report contains no explicit statement of the researchers' hypothesis, though the first sentence of the penultimate paragraph of the introduction does suggest it:

> … students who were taught from the slides with sentence headlines [will recall] … significantly more information than those students who were taught from slides with phrase, question, or no headlines ….

> Of course, this implicit hypothesis is not stated in null form. The null version of the hypothesis would be

> There will be no statistically significant increase between the retention of students taught from slides with sentence headlines and the retention of students taught from slides with phrase, question, or no headlines.

The fact that the null hypothesis is not explicitly stated is not a significant flaw in the article; one might even argue that including the formal hypothesis statement might have hurt the tone and readability of the article for its intended audience. And it is most unlikely that a reasonably intelligent reader would fail to infer it from the context. Still, there would be no question that readers would understand exactly what the authors were testing if they had stated their hypothesis explicitly.

Test. The hypothesis test is effectively summarized in Figure 8.1. Two sections of students were taught using slides with phrase, question, or no headings; two other sections were taught the same material by the same professor using slides with succinct sentence headings. After the lectures, they were provided with copies of the slides to use as study aids. An examination assessed their learning of material presented in the slides.

Because the examination consisted of computer-scored multiple-choice questions and because each question could be related to information presented in a specific slide in either the original or transformed version, Alley and his coauthors were able to compare the answers across the control and experimental groups, and to assess their learning based on the differences in headlines in the slides presented to each group. They were also able to compare answers to 40 control questions taken directly from previous examinations in the course (Table 8.1). Students prepared for these questions by studying the earlier examinations rather than studying slides, so these questions provided an assessment of the overall effort in preparing for the examination for each section of the course.

In their Experimental Methods section, the researchers provide explanations of the following:

- How they transformed the slides
- Why they chose the specific design (type size and style, and the use of graphics rather than bullet points in the body of the slide) used for the slides presented to the control group

- Why the slides were the major difference between the four sections (two for the control group and two for the experimental group), despite the fact that the lectures involved consisted of four different spoken "performances" of the same information

Exercise 8.1

Using the same null hypothesis used by Alley and colleagues, suggest a different way of testing their hypothesis. ■

Sample Selection

You will recall that we talked about samples of convenience in Chapter 4—populations to which the researchers have ready access. A sample of convenience (in this case, the students in four sections of a geoscience course) compromises the principle that every member of the general population should have an equal chance of being randomly selected for inclusion in the study. As a result, the researchers must either demonstrate how the sample is still representative of the general population or at least qualify their findings with appropriate limitations.

In the first paragraph of the Experimental Methods section, the authors demonstrate that their sample is representative of the general population of undergraduates at Virginia Tech (more than 20,000 students). They note that "Because the course satisfied one of the university's general education requirements, it was a popular course for nonmajors, attracting students from all branches of science and engineering, as well as those from liberal arts, agriculture, and business." Moreover, the sample size is very large. The control group consisted of two sections of 200 and 202 students, while the experimental group consisted of two sections of 201 and 136 students.

Alley and his colleagues do not impose any limitations on the generalizability of their findings, however. In fact, we could argue that they do just the opposite. Because the statistical significance of the difference in the mean percentage correct in the answers to the fifteen questions is so great—1 chance in 1000 that the results are due to coincidence (Table 8.2)—and because the situation studied with the sample population is so similar to the general situation in which slides are used in scientific and engineering presentations at conferences or in industry, the authors feel confident to state in their Conclusions section that "if technical presenters desire to emphasize assertions in a presentation, they would do well to place those key assertions in succinct sentence headlines."

Exercise 8.2

Assume that you want to replicate the research that Alley and colleagues report in this article using a professional rather than a student setting. How would you change their study to accommodate this change in population? ■

Report of Results

The results of the study are reported primarily in Table 8.1 (average score for each of the four sections on the 40 control questions taken from previous examinations in the course) and Table 8.2 (percentage correct for the control group and experimental group for each of fifteen questions based on the slides that had been transformed for the experimental group. Table 8.2 also reports the significance level of statistical difference in percentage correct between the control and experimental groups.

We can see from Table 8.1 that the experimental group not only did not put forth more effort in preparing for the examination, but in fact scored lower on these questions. No statistical analysis of the differences among these mean scores is provided, however.

Table 8.2 shows that for seven of the fifteen questions drawn from the traditional slides presented to the control group and the transformed slides shown to the experimental group, the experimental group performed better, with a very high degree of statistical significance. For another six questions, there was no statistically significant difference in learning. For two of the questions, the control group scored higher, with a very high degree of statistical significance. Finally, the mean scores of the two groups show that the experimental group scored significantly higher overall on the fifteen questions, again with a very high degree of statistical significance.

Analysis

Based on the results they report, Alley and his colleagues conclude that audience retention is increased to a very high degree of statistical significance when presentation slides use succinct sentence headlines rather than phrase, question, or no heading. We need to examine their analysis to ensure that it meets the standards of rigor for inferences based on measurements: The validity of the measurement and the reliability of the inference.

Validity. You will recall from Chapter 4 that there are two types of validity: internal and external. Internal validity in a quantitative study addresses the question, "Did you measure the concept you wanted to study?" The concept that Alley and his coauthors wanted to study was the difference in audience retention of information presented in two different types of presentation slides. The key to determining the internal validity of their study is to examine how they operationalized the concept of audience retention.

Retention is a concept that we are all familiar with, whether we are currently enrolled in a course or not. Simply put, retention is the ability to recall information, and we can measure that recall by presenting audience members with questions that test their ability to remember specific information. Because the study was carefully designed, the concept was validly operationalized, and the questions on the examination tested the students' ability to recall information presented in the slides, it is clear that Alley and colleagues did measure the concept they wanted to study.

External validity answers the question, "Did what you measured in the test environment reflect what would be found in the real world?" We can manage external validity by taking care when we design our test that the conditions in the test environment match the conditions in the general environment as much as possible and that the sample group itself is a fair representation of the general population of interest.

Alley and his coauthors explain briefly how their experiment involving students and lectures in a university course differs from "real-world" scientific and engineering presentations attended by technical professionals and managers. Paragraph 7 in the Introduction points to both similarities and differences between the two situations and audiences. A fuller explanation might have been helpful, but it is clear that the intended audience for this article understands those similarities and differences.

We looked earlier at the sample population for this study, and we remember from Chapter 4 that subject selection must be random; that is, the selection and assignment of test subjects should be independent and equal. Selection is independent if selecting a subject to be a member of one group does not influence the selection of another subject. Selection is equal if every member of a population has an equal chance of being selected.

The test population for Alley and his colleagues was clearly a sample of convenience that was not selected by the researchers. Indeed, to some extent, the population was self-selected (after all, students enroll in a particular section of a specific course for a variety of personal and practical reasons), but there is no evidence that students' decision to enroll in these sections or not was related to the study because there is no indication that the students were aware that the study would be conducted when they enrolled in the course.

We can fairly say, however, that the self-selection of the 739 students who participated in this study by deciding to enroll in these sections of this course was both independent and equal.

Reliability. Again, you will recall from Chapter 4 that the reliability of a study involves the likelihood that the results would be the same if the experiment were repeated, either with a different sample or by different researchers. The measure of the study's reliability is the statistical significance of the results. In other words, how confident can we be that the difference in performance between the control group and the experimental group is due to the intervention (the independent variable) introduced rather than to coincidence.

In inferential statistics, we judge the reliability of our conclusions about a set of data based on two principles:

- The smaller the variance in the data, the more reliable the inference.
- The bigger the sample size, the more reliable the inference.

When we examine the variance in data, we want to know the probability that the results were caused by the intervention and not by sampling error or coincidence. You will recall that in Chapter 4 we observed that most research in our field is considered rigorous enough when results have a p value of 0.1 or lower; in other words, there is no more than 1 chance in 10 that the results were caused by coincidence or sampling error. For more conservative situations where harm could be done by accepting a false claim, then a p value of less than 0.05 or even 0.01 might be more appropriate; in other words, there are no more than 5 chances in 100 or 1 chance in 100 that the difference is not statistically significant.

The results that Alley and his coauthors report are really amazing because of the very high degree of statistical significance. For three of the fifteen questions that tested the effect of the intervention, the experimental group scored significantly higher than the control group at a 0.001 level of significance—that is, 1 chance in 1000 that the difference was the result of sampling error or coincidence. For another three questions, the experimental group scored significantly higher than the control group at a 0.005 significance level, and for a seventh question, the experimental group scored significantly higher than the control group at a 0.025 significance level. It is exceptionally rare to find such low variance in quantitative studies in technical communication.

If we can fault Alley and his coauthors at all, it is that they do not adequately address the other eight questions for which there was either no statistical difference in the percentage correct between the experimental group and the control group (six questions) or for which the control group scored higher at a 0.01 level of significance (two questions). Although we might call the inferior performance of the experimental group on the last two questions exceptions (no test, after all, is perfect), it is more difficult to explain the other six questions for which there was no statistically significant difference, though on all but one of those questions the experimental group did score higher than the control group.

Despite this problem, however, we must conclude that the degree of variance on the first seven questions was so low as to assure us of the reliability of the inference that the use of sentence headlines can increase audience retention.

As for sample size, the experimental and control groups comprised 739 students out of an undergraduate student body of approximately 20,000, with a control group of 402 students and an experimental group of 337 students. This sample is significantly larger than is ordinarily used in classroom studies conducted in technical communication research. As a result, we can conclude that the size of the sample also increases our level of confidence in the reliability of the inference that the use of sentence headlines can increase audience retention.

Conclusions

The final section of the article is brief but effective. The authors first draw on data from Microsoft estimating that 30 million PowerPoint presentations are made every day. Although the authors do not make an explicit connection between this observation and their conclusion that sentence headlines significantly increase retention, they certainly suggest that retention of the content in these 30 million daily presentations would be increased if presenters used sentence headlines in their slides.

The authors then address the problem that PowerPoint defaults make it difficult for presenters to implement sentence headlines, and they point to templates that will remedy this problem.

Finally, the researchers note that this pilot study should be followed by additional testing to determine the influence of other differences in the design between the original and transformed slides used in their research, particularly typography and the use of graphics rather than bullet points in the slide's body.

What is particularly notable about the conclusion of this research report is the rather low-key assertion of its significance. Although leading off the conclusion by suggesting that 30 million presentations each day could be improved if the slides used sentence headlines certainly makes a rather grand point by implication, the conclusion of the study itself is modest:

> In our study, using sentence headlines of no more than two lines led to statistically significant increases in recall from the audience on details that were contained in those sentence headlines.

Summary

This chapter has expanded on the concepts and methods presented in Chapter 4 by examining an example of a quantitative research report. Following the full text of the sample article, we have explored its purpose and audience, its structure, the study design, the results, and the inferences drawn from those results. We have seen that the quantitative study conducted by Alley and his coauthors meets the requirements of internal validity because it operationalized the concept it wanted to examine. The research project also meets the requirements of external validity because the test environment, including the sample population studied, was sufficiently like the "real-world" environment. Finally, the inferences that the authors draw from their results are reliable because of the very low probability of error and the size of the population sampled.

References

Alley, M., M. Schreiber, K. Ramsdell, and J. Muffo. 2006. How the design of headlines in presentation slides affects audience retention. *Technical Communication* 53:225–234.

Answer Key

Exercise 8.1

The answer to this exercise will be unique for each person who prepares it, so there is no key to this exercise.

Exercise 8.2

The answer to this exercise will be unique for each person who prepares it, so there is no key to this exercise.

9

Analyzing a Qualitative Research Report

Introduction

In Chapter 5, we discussed qualitative research methods such as interviews, focus groups, usability tests, field observations, and document analysis, techniques that draw on research methods frequently used in the social sciences. We noted that these qualitative methods typically involve three phases: observing behavior, recording data, and analyzing the data. We defined the standards of rigor and ways of ensuring rigor in qualitative studies, and we also described coding and categorization schemes used to record and analyze qualitative data. In this chapter, we will examine a sample article reporting the results of qualitative research. The chapter contains the full text of Greg Wilson and Julie Dyke Ford's "The Big Chill: Seven Technical Communicators Talk Ten Years After Their Master's Program," which appeared in the May 2003 issue of *Technical Communication*, as well as a detailed commentary on the article.

Learning Objectives

After you have read this chapter, you should be able to

- Analyze a qualitative research report
- Apply the results of your analysis to reporting a qualitative study on your topic of interest

The Article's Context

Because qualitative research methods are so commonly used in the social sciences, it's probably not surprising that Wilson and Ford's article "The Big Chill" is essentially a sociological study of seven young technical communicators. It recounts their experiences on the job and the ways their views of the profession and roles within it have changed over 10 years. It also investigates the wide range of work that technical communicators perform in business, industry, and academe, as well as these young professionals' satisfaction with the work they do.

The article is significant for several reasons. It is one of a relatively small number of qualitative research reports in our field and one of an even smaller subset that uses a focus group as its data-gathering technique. And "The Big Chill" is particularly instructive as an exemplar of focus-group research because it includes an essentially complete transcript of the discussion, thus giving the reader significant insight into how the researchers analyzed and interpreted the data.

The article is timely because it reflects some of the principal inward and outward focuses of the profession. It examines the extent to which the communicators' academic training prepared them for the realities of the workplace. It also attempts to probe the reasons for burnout among technical communicators in their early working years, as well as attitudes of colleagues in other disciplines toward our profession and those who practice it.

In other words, "The Big Chill" is not only a really interesting example of an important type of research report, but it is also a really interesting article.

"The Big Chill: Seven Technical Communicators Talk Ten Years After Their Master's Program"*

Greg Wilson and Julie Dyke Ford

Introduction

We begin this article with the edited transcript of a listserv conversation among seven individuals who graduated in 1991 from the same master's program in technical communication. The purpose of the listserv was to generate discussion among individuals who have been in the technical communication field for over a decade. As researchers, we were particularly interested in knowing how the educational training of these individuals influenced their acclimation to the work world. We were also interested in understanding the roles they played within their organizations, as well as comprehending the ways in which they interacted with coworkers, and identifying the triggers for feelings of burnout and disillusionment that many of the subjects experienced.

The experiences and opinions expressed in the listserv conversation pertain to valuable issues in our field—issues that are of interest to both technical communication educators and practitioners. Through their candid discussion, the listserv participants address the strengths and weaknesses they see in their training, their jobs, and the profession as a whole. In the analysis section of this article, we draw on the frank comments of the participants as a way to begin to understand the challenges that technical communicators face, both as they are adjusting to organizations and even when they have established themselves in their careers.

In highlighting reported differences between expectations and reality, we hope to provide insights for students preparing for careers in the field and for the professors who instruct them. In discussing the problems encountered (and roles played) by working professionals, we hope to point to issues that the profession should address and to help individual technical communicators realize that the problems they face are shared by others in the profession. And finally, as we address burnout, we hope to call attention to some of the structural issues (that is, the workplace expectations and realities that make success and fulfillment elusive for many technical communicators) that drive many talented people from the profession.

* This article was originally published in *Technical Communication* 50:145–159. References' and citations' styles have been changed to conform to *The Chicago Manual of Style*, 15th edition. Reprinted with the permission of the Society for Technical Communication.

Methods

We solicited discussion participants by sending a request to an alumni discussion list for a technical writing master's program. We requested participants who had graduated in the 1991–1992 time frame. We then set up a listserv that included the seven volunteers and ourselves. The listserv discussion took place between 15 October 1999 and 15 March 2000.

We began the discussion by asking each participant to write a professional narrative that described what they had been doing since graduation. Even though some of the classmates had kept in touch, not everyone knew the history or status of the other people on the list. The two authors of this article also contributed professional narratives so that the participants would be more comfortable with those who were asking the questions.

Once everyone had contributed a narrative, we posted the first question to the list and asked the participants to post an answer to the question and to discuss and react to the comments made by the other participants. When the traffic on the list died down, we posted a new question to the list. We did not attempt to moderate or contribute to the list; we would simply ask questions and occasionally send out reminder notes to the group when it seemed that they were not responding to questions (for example, over the holidays).

The first two questions on expectations vs. reality and lessons learned were planned before the discussion began. The questions on burnout and roles grew out of comments and themes in the ongoing discussion.

The text of the discussion presented in this article has been edited for length and to remove off-topic discussions and references to specific people and organizations. The text we include is substantial because we feel that the extended interaction of the participants brings issues to light that would not be addressed well simply by excerpting sentences or phrases. In our analysis of the discussion, we have treated it as an interactive autoethnography—a critical examination of the culture of technical communication by technical communicators that pulls together their perceptions of cultural patterns in a way that only those with membership in that community could.

One could question how representative this conversation and its occasional criticisms are of technical writers as a whole. It was not our intention to do a scientific survey of attitudes across the profession—many studies of that type have been done. Our goal was to allow the voices of professionals to relate their experiences, and one must trade off breadth of coverage when seeking that kind of depth.

From a similar starting point, however, the seven participants in the listserv have been involved in many different professional situations (for example, engineering firms; dot-com startups; education; consulting; software, hardware, and biomedical firms) in companies of various sizes, and are able to discuss a wide range issues pertinent to our field. And as graduates of a master's-level technical communication program, arguably they are representative of professionals who enter practice in the field with the most preparation and training, representative of the types of students current degree programs are training.

As such, they may be somewhat unrepresentative because they were better prepared to deal with the problems they discuss than many who enter the field, but still, there is much to learn from their comments.

Discussion participants. Alan Sloan is a documentation manager for a health-care software company in Massachusetts.

Tom Kolano is a knowledge management consultant for an international software firm.

Jill Ann Mortell teaches technical writing to undergraduates and confines her writing to poetry and reviews of poetry while pursuing an MFA.

Monica Mitchell is an adjunct instructor of technical writing and a university counselor, and she has research interests in screen and document design.

Mara Cohen Ioannides teaches composition at a large midwestern state university and has research interests in technical writing and rhetorical studies of American Jewish documents.

Bill Pollak is a manager of communications at a software research institute, overseeing a staff responsible for technical communication, communication design, public relations, and Web development.

Margo Parent specializes in writing end-user documentation at the programming and system administration level for various software products.

Background of participants. The master's program where these participants were enrolled is at a prestigious U.S. university. They were drawn to the field for the various reasons any of us are: a fascination with language and technology, a desire to find a creative job that pays well. The students took three semesters of coursework, and an internship was required after the first year. The program was modeled on a professional degree program, so no thesis was required.

Students took courses in areas including technical and professional writing, graphic design, desktop publishing, rhetoric, computer science, research methods, and organizational management. The curriculum of the core courses in technical communication was very project driven, and professors brought their experiences working within industry into the classroom.

The students were genuinely excited about the program and the work that would follow their graduation. They were going to go out into the world and solve communication problems, make a good living, engage in stimulating and interesting tasks, and work in exciting environments. Like the group of 1960s college students in the film to which this article's title alludes, the enthusiasm and expectations of this group of technical communicators didn't fully turn out the way they expected. Many have switched companies several times. A few have left the field. And a few are still very excited about the jobs they hold and the changes those jobs have brought to their lives.

Listserv Discussion

Q *When you entered the tech writing field, what were your expectations in terms of the tech writing profession, the industry in which you were about to work, and the ways in which your education would help you? How well did those expectations mesh with your experience on the job?*

Mara *I felt that our MA program prepared us very well. Of course, I had also interned at a software institute and shared an office with Bill. I think that experience was great and the let-down was my other jobs. The institute offered a potpourri of writing jobs and that, for me, was thrilling. But my other industry jobs were rather one sided, and after 6 months I knew that long term was not for me.*

I like having a number of projects happening all at once, and working on just software or just hardware I found to be too confining. However, that is my problem. I also found working with engineers whose lives were about their work to be difficult. I'm a more gotta-get-home-and-make-dinner kinda gal. I guess I expected the jobs to be like my internship and not like they were.

Margo *One expectation I had on entering the tech writing field was that our profession would be embraced by other fields (other departments in a company, for example). I learned very quickly that this is not always true. I have had the distinct displeasure of working with some very difficult subject matter experts (SMEs).*

Almost immediately, with my internship, I was hit by how quickly a writer can be treated as a second-class citizen. Only once during my last four jobs have I been truly embraced by the engineering team, which is where I seem to end up reporting. They were grateful for my talents, and "used" me wisely. This has led me to a conclusion: "The reality of dealing with SMEs" should be in the curriculum for any professional writing program.

Other expectations I had: I expected that my MA would secure my position in a rapidly growing field where every other person is obtaining a certificate from "Joe's College of Writing," and I expected more money. Both of these have proven true for the most part.

I would agree with Mara on the point that the program prepared us well for a variety of writing projects. I would say, however, that I do not think they prepared us well for the "real world" (as they always liked to say). Courses on how to deal with difficult people, showing how your efforts as a writer reduce costs in other areas of a business, budgeting documentation projects, etc. Some of this is learned through an internship, but not always.

Bill *I entered the tech writing field with some specific goals in mind. I had spent most of my adult life in the music business, and I was looking for a career that would provide me with stable income doing work that was intellectually challenging and stimulating. I expected my education to provide some validation of my skills as a writer so that I would be able to get started in this profession.*

I also expected my education to teach me about the business of professional writing so that I could enter the job market as something more than a bumpkin who thought he could write a little bit. Today my job meets, and maybe exceeds, those expectations. I'm happy with the income, I've learned a lot about things I never would have had the curiosity to investigate on my own, and I am intellectually challenged and stimulated every day. A positive unintended consequence has been that I am able to pursue music with much more freedom than I ever thought would be possible.

Tom *I suppose I was more idealistic.... I expected there to be a lot more creativity involved in a tech writing job. Again I think I had some bad luck; there are much better tech writer jobs out there where you are respected in a truly collaborative environment, instead of reduced to a grammar checker. My experience instilled a fear of being pigeonholed and encouraged me to get a job with broader responsibilities.*

My experience jibes with Margo's in that sometimes you can be considered a second-class citizen. Sometimes you're thrown into an environment where a manager thinks they need a writer but lacks an understanding of what you can and cannot do, and the vision for that position doesn't end up meeting the original plan.

I thought our program would prepare me very well for the profession, and (speaking only for me) it was a perfect match. The multidisciplinary nature of the degree was key: the design aspect, desktop publishing, cognitive preparation, and argumentation being my favorite slice of the pie. I would describe it as, the degree gives you the benefit of the doubt in the beginning as far as respect, and after that you prove it right or wrong with your own actions.

Margo *Tom makes a few good points.... I have found that some employers hire a writer because they think they have to—it's the thing to do—not because they are aware of what the writer can do for them. (Not once in my career have I reported to someone who at some point was a writer, or knew anything about it.) So, I have found, I end up educating the company. But I have been lucky in that I have been able to make a job what it needs to be for myself to feel accomplished and satisfied.*

Jill *In many ways, the degree prepared me well. I think that's mostly because I had internships with a software institute, the robotics group, and computing services while I was still earning the degree and just after I completed it. I knew what types of writing to expect. What surprised me in the real world were a strange conglomeration of things.*

1. *As a freelancer, I was able to earn more right away than I thought possible. This happened on a few Department of Defense-related jobs with deep pockets and a steady gig near the end where I was basically on salary to be available to one company.*
2. *I spent a lot of time not writing—my least favorite pastime became those meetings full of people who love to hear themselves talk and who never come to any decisions. I also spent a lot of time teaching and hand-holding on the job (with people in positions that made me believe they should know this stuff).*
3. *I was often one of very few women on a project (if not the only woman). I found dealing with departments full of male engineers with Department of Defense/military leanings to be very difficult. Often I was treated as a glorified secretary by people I was trying to collect information from, and consequently I had to get very in-your-face, which is not my style. It was very isolating to be a contract employee in that type of environment. I also never anticipated how difficult it would be to manage other writers and graphic designers on large projects. I assumed that other writers would be fairly organized, would need little prodding, and would generally act like adults. (Wrong!)*
4. *It was very easy to end up doing the same types of projects over and over again, even as a freelancer. The repetition, though easy, was boring to the point of being numbing.*

When I started working full time I had no sense at all how easy it would be to get burned out. I assumed I would be able to do many of my freelance jobs from home, but any job of substance took me on site for long hours. At the time, I was a single mother of three and I would come home at night and—while my briefcase was still in my hand—the babysitter would be telling me about school projects and Boy Scout picnics and who got a D in English, and I would want to die. I finally thought there must be an easier way for me and my kids to live, even if I had to walk away from the money and from some clients I really liked.

Mara *Tom and Margo raise a problem I had, but differently. The first company I worked for knew what a writer was but expected me to do the impossible because they thought I was psychic. Create documentation for software that doesn't yet exist except in meetings. And then when the documentation didn't appear, I was the one under fire, even though the engineers might never meet their deadlines. I found, even in the mega company, the writers to be far less petty than the engineers—and better organized.*

Monica *Since computers would inevitably serve an ever-increasing role in business and industry, I desired a tech writing program that would give me enough new information and new technology to transport me to a level or more above the general knowledge base of most users in business and industry. Our graduate program did just that. Even now, nearly 10 years later, the document design and cognitive psychology studies continue to be very useful in my work.*

Q *Since graduating from your tech writing program and entering various different corporate/academic environments, what are the most important things that you have learned in your jobs as technical communicators or about the profession as a whole?*

Mara *I think a great generalization is to say that being a technical writer is like being a horse on* Animal Farm. *Everyone knows we are necessary, but we aren't as "good" as the pigs (that is, the engineers). Not to say that engineers are pigs. I often felt that I was a necessary evil, part of the team but not really. I'm still coming to realize that people in general don't realize the importance of clear communication but understand bad communication.*

Monica *One of the most important things that I have learned as a technical communicator is that corporations and businesses still have a lot of catching up to do. As I mentioned earlier, I have found that most companies are full of "computer users stuck in the typewriter era." On graduation in 1992, I found the "lag" if you will, the purposeful lack of curiosity as to how manipulation of text (thanks to computers) might easily enhance the readability of a page, everywhere. And now, nearly 10 years later, what do I find? Employers/employees using computers to produce pages that look like they came from a 1950s Royal Deluxe.*

Margo *I think I would agree with Mara—I like her analogy about the horses and pigs in* Animal Farm. *Also, it has become more and more evident how we as writers rely so heavily on others to get our work done—and that this is not always a fun aspect of the job. A manager in one company I worked in wanted me to write a support guide, yet he continually and consistently withheld pertinent information, even after he had been asked to surrender it. Actually, I think this was a character flaw, but it nonetheless illustrates my point—I was deadlocked until I somehow coaxed it from him, or worse yet got my boss to do it.*

I am not sure we were given a fair assessment of the "real world" in graduate school. I have learned that as a technical writer you have to: be investigative, almost sneaky sometimes; play politics; be prepared to hit brick wall, after brick wall, after brick wall; and have the stamina to break down or climb over the walls. These tasks can be very tedious when all you really want to do is write and bring clear communication to your customer.

Mara *In response to Margo (and others) who want other training in tech writing programs, let me speak from the teacher's view. Other than internships, I'm not sure how one would incorporate such training into a program. Remember that management course we all slept through (I think I still have the book in my basement)? That was not helpful.*

I understand the need to teach management since we either end up leading, or "sneaking," but how would one teach politicking in a class? Especially when the basics (grammar, design, graphics, etc.) need to be beaten into the students. I wish we had been taught not only how to create our own schedules for projects, but how to encourage others to meet our deadlines and how to incorporate others into our schedules.

Margo *I think Mara is correct in that it is difficult to teach politicking per se. But a good general business course wouldn't hurt (not a management course)—how a business is commonly set up, where the tech writer usually reports, what conflicts can arise from this, how to deal with this (that is, how to educate your fellow workers on your role and important contribution to the team), how to deal with difficult people. But in the end, maybe it boils down to personality. I am fortunate that I have the backbone to not put up with any B.S. But I have seen some writers, and other coworkers, completely steam rolled because of their lack of business and people skills in the work place.*

Alan *I think that the curriculum we had in our graduate program has been very useful in my work as a technical writer. I do wish that, as Margo said, we had been taught interviewing (and interrogation) skills. I've learned that you have to give technical experts a reason to want to give information. I was naive coming out of school thinking they would do it just because it's their job. My success at getting information has come from establishing relationships with them on personal and professional levels. It's come from being friendly with them at the water cooler, I've scanned a photo of their kid for them, or I've stroked their egos by asking them to explain a technical concept (regardless of whether I already understood it or not).*

As for the management class, I wish I had paid attention and gotten more out of it. Coming out of the program, with the experience of my contract and my coursework, I was prepared to be a good technical writer. I was not prepared to deal with the environment of corporate America. I was not prepared to be a self-managed department responsible for estimating multiple, simultaneous project schedules with already-set deadlines.

I like Margo's idea of some course on business fundamentals rather than the management course. It would have been useful to have a basic understanding of a company's financials so I could have known before accepting a new position that I was jumping onto a drowning ship. It would also have been helpful to have a basic understanding of the product life cycles, project management, the differences between an engineering-driven and marketing-driven company, and things like the pros and cons of working in marketing vs. engineering vs. QA departments.

Perhaps the business courses should be taught as part of a concentration for students who want to pursue management or consulting career paths. Although I know that when I left graduate school, I did not see myself as going down the management path as soon as I did.

Jill *I agree with Mara and Margo. I find that being a writer in any organization I've worked in (and I've experienced many organizations as a freelancer) means also wearing hats that have nothing to do with writing. I've filled the role of investigator, interrogator, hand-holder, secretary, gopher, manager, you name it.*

One of the things I liked about writing before *I entered the profession was that it's a pretty solitary endeavor: you collect the information with as little ado as possible and then go home and write—or so I thought. Not so. The politics and coaxing information from people, etc., got old fast.*

I also found that the extraordinarily fast pace of technological change was difficult to handle as a freelancer. If you specialize or work for one company, I think you have a shot at staying current in whatever your area happens to be. As a freelancer, I was expected to know any and all software and hardware at whatever organization I was working with on a particular day, as well as to understand or at least have a rudimentary grasp of tasks and concepts that were not my job but were pertinent to it (such as object-oriented programming).

Q *Several people have mentioned or alluded to the issue of burnout, either as an impetus to change jobs or a reason to leave the profession. What have been your personal experiences with the issue of burnout? Do you think this is a problem of particular concern in the tech comm profession? Why? Has burnout been a factor in your career path? How do you fight it? When do you know when it is time to give in? And then what?*

Margo *Only recently have I felt burned out. This may be due to the fact that I have worked for 3 startups in the last 7 years. I do not think that this problem is common only in the technical writing field—I think it is a product of our culture. We are so bent on being "successful" and making a lot of money, that we are driven to work crazy hours in fast-paced industries. We all fall into the "trap."*

I don't know how to fight burnout. I am currently having a very difficult time with it. I find I have to take things day-to-day and reassure myself that I don't need to remain in this job/profession forever. I think the trick is deciding whether you really want to leave it, and then what you want to do instead (if you don't happen to be independently wealthy).

Jill *I think Margo is right to some extent in her assessment that burnout is a product of our culture. I also think, though, that our particular profession is prone to burnout.*

I began to feel really burned out when I realized what little difference I was making for the companies I worked with, the reader, and society in general. This is ultimately what leads to the demoralization that is commonly referred to as burnout. The rest is just exhaustion. I finally felt that I had to find something to do that felt more useful and also fed me.

This is not to say that technical writing isn't useful, because it is. But the nature of the profession is a lot of wheel-spinning, a lot of things that have little to do with the writing itself. And the writing (let's face it) is such a small part of what most of these companies are doing. Most end-users don't even read the manual (or the whole journal article, or the annual report, or whatever).

Now, in my teaching job, I try to look at what I do as a semester's worth of insurance for my students. I tell them what they learn in my classroom is insurance against ulcers and wasted time when they get out there on the job. So that one day, when their boss strolls into their cubicle and says, "write me a report about this," they can feel confident in performing the task at hand efficiently and successfully. I have to look at it in those immediate kinds of terms. I can't think about what it means to the reader or the company. I know better.

Mara *I guess some bad jobs started me on burnout earlier. I felt used quickly, and my personal philosophy that a job is a job, not a lifestyle, upset many of the programmers. Since my first 18 months I've decided I wouldn't return to industry unless I went as a freelancer or negotiated my hours to my liking. Besides I'm really not an 8:00 to 6:00 kinda gal.*

That's not to say academia doesn't cause burnout. I am a bit burned out. I teach the same things over and over again to students who don't care/aren't interested/etc. Although I've had some wonderful ones too. However, my schedule is such that my interests are my research, and if I don't teach or have office hours, I can go home/shop/read/ etc.

There are weeks that are slow, and in industry I found those to be painful because I was stuck at the office treading water, but here I can do other things to plan for the hell weeks, like where I'll have 80 papers to grade in 5 days. I think a lot of the industry burnout is related to boilerplating and unappreciative coworkers. Like Jill, I try desperately to explain to my students that I'm trying to make them better able to operate in the business community.

Alan *For me, the feeling of burnout has resulted from what we discussed earlier about knowing when to fight, when to bang your head against the brick wall, and when to walk away. Every place I have worked I felt that I made tremendous progress on educating engineers and management on the value of technical writers. Then, when the time came that I felt I was no longer making a difference, I began to feel the burn. More than once I felt as though I was about to lose all the ground that I had gained, so I left while I still had some feeling of accomplishment.*

I believe that burnout is an issue for any technical writer who has a passion for what he/she does. When you stop caring, it's time to get out. Having experienced it myself, as a manager of tech writers, I have found that it's easy to recognize the signs. From the manager's perspective I see a couple of ways to prevent it or at least alleviate it, providing it has not gone too far:

1. ***Customer visits*** *Let tech writers meet the people who read and use their material. I've had very positive experiences with this method. It reconnects writers with their audiences and the environment in which they use the product.*

2. ***Training seminars*** *Send writers to courses on new applications or fact-finding seminars to explore new technology. Sending them to stress-reduction seminars is a sure way to tell them you have nothing better for them to do.*
3. ***Special projects*** *I've given tech writers who normally write user guides the opportunity to help manufacturing personnel write ISO-9000 procedures for their work areas. I've also had them work with marketing communication to write user stories or application notes. It puts a bit of variety into their job and gives them a sense of connection with other parts of the company. It also increases their visibility and perceived value.*
4. ***Professional associations*** *Encourage writers to participate in STC, SDP, IEEE, ACM, etc. Send them to conferences or training or even encourage them to write a journal article.*
5. ***Continuing education*** *Encourage writers to take advantage of corporate benefit packages that offer educational reimbursement.*
6. ***Incentive plans*** *Fight for tech writers to receive bonuses so that they are motivated to meet goals just like sales people and lead engineers.*
7. ***Communication*** *Make sure that tech writers know what's happening within the company in a timely fashion so that rumors don't distract or demotivate them.*
8. ***Recognition*** *Make sure that tech writers are recognized when a product development team receives recognition at a company meeting.*

All these things are ways to demonstrate to tech writers that they are valued for what they do and that their manager appreciates that they need to grow professionally. The problem with being a solo tech writer (or a tech pubs manager) is that you have to educate your boss on how these things work and then be able to recognize for yourself when to tell him/her that you need one of them.

Q *Another issue that has come up repeatedly is how companies/employers often don't know what to do with a tech writer or how to use them. Could you all elaborate on your experience with this issue? Why are employers lost when it comes to effectively integrating technical communication professionals? Is there a problem in the way they think about products/employees/information? Does the problem relate to the technical communicator's need to span boundaries? What are your suggestions for better ways to educate employers and technical communicators to resolve this issue?*

Margo *Documentation is still, for the most part, an afterthought. Companies do not embrace it as part of "the product"; they do not include it in the plan. Why is not exactly clear—it seems obvious that to complete a product you need to write the code, test the code, and create documentation/deliver information about the product so that people can use it effectively.*

Unless an organization or individuals in an organization have worked with an experienced writer, they do not know that a good writer can do more than edit some engineer's garble. They are unaware that the good writer can be the advocate of the user, and design and deliver information that makes the user's job easier. It doesn't occur to them that if the user finds the answer in the documentation, they will not call support. And when they do not call support, some resources have been saved.

In addition, so many untrained professionals are doing technical writing and getting by. So why would a company that employs such an individual ever see the writer as more than a glorified secretary? And why would that company give that individual any authority in doing anything but word processing?

I think it boils down to the company's beliefs and experiences. These things are hard to counteract, but I think the best way is to just do—prove one's skills go beyond those of the typist, make suggestions about how those skills can carry over into other areas. Also, present the business cycle I presented above—most managers will respond to the bottom line; everything is about money. Hopefully, they begin to see the value. If not, it may not be worth the fight, and it's best to move on.

Mara *I'd like to respond to Margo, or rather build on her comment. I think the problem is older than company history.*

As a writing instructor, I've come to the conclusion that writing is not respected. Thus, all those developers leave college believing that writing is not important, an attitude that carries over into their work experience. If they didn't like writing, couldn't get it down effectively, and had bad experiences in writing classes, then you cannot expect them to respect writers in the workplace. Add that to a poor company expectation, and writers have a poor reputation.

So we have to go back to the education system and convince people that good communication is not only important/necessary, but not as difficult nor as simple as they would think. Not everyone can do complex communication effectively, just as not everyone can do complex programming effectively.

Alan *I agree with both Margo and Mara in their assessments that, in general, people and companies do not place a tremendous amount of value on writing. This is especially true for user docs. I wish I had a nickel for every time some engineer or marketing know-it-all has told me that "no one ever reads the manual" or that "we're just including manuals to minimize our liability." The majority of engineers and marketing staff believe that "anyone can write"—an oversimplification that results in a diminished perception of the value tech writers bring to the product development process.*

Again I could retire at 35 if I had a share of Microsoft for every engineer that assumed that I'm just a frustrated engineer "who could not handle the math." Very few people, if any, believe that I chose technical writing as a career and went to school for it. Most of the writers I've met over the age of 35 found their way into the field through either frustration with book publishing or as a way to move out of the secretarial pool or the manufacturing floor without an engineering degree or MBA. The few who have degrees have earned certificates as part of a career change after being laid off by big companies in the late 1980s and early 1990s.

I think there are a couple of avenues to change attitudes. To change tomorrow's workforce, we have to pursue changes within the educational system. At a basic level, if we placed more emphasis on professional written communica-

tion skills in high school and college, especially in the technical fields, there might be an increase in the awareness if not respect that people have for professional communicators.

To change today's leaders-in-training, we need to implement change in both engineering and business schools that teach the product development cycle. If they included the role of technical writers in the discussion, it could mean increased awareness.

And finally, I know this seems a bit radical, but I think we should have a professional licensing exam on the same level as a professional engineer license. My degree and ability to discuss the many facets of technical communication work to a certain point, but having a standardized, cross-industry recognized license would be that much better. It would provide a professional validation and help the uninformed hiring manager identify trained professionals.

Mara *Let me preface this with the note that I'm in the midst of grading 80 final papers and rewrites, so this may be harsh/biased/angry.*

The education system needs to be revamped without a doubt, and like Alan, I agree it must begin earlier than university. I am awfully tired of having to remind my students that the number one requirement of employers is not a degree in the field of employment but the ability to communicate effectively, both orally and textually. I'm awfully tired of students who rank a writing class at the bottom of their priorities because writing will not affect them in the future because they won't have to do it.

Jill *I don't know why this should be the case, but my experience has been that the larger the company, the more clueless it is about how to use a technical writer. Because I've freelanced for several years, I've worked with many different types and sizes of companies. Obviously, the company I worked for that specialized in document design and subcontracted me to write user manuals knew how to use me. They let me deal with the client entirely and gave me free rein in writing and design, and even took my suggestions about who should illustrate the manuals. The mom-and-pop high tech businesses and small start-ups looked to me to solve their (considerable) problems and always treated me with respect and took my advice and at least behaved as if they were grateful for it.*

It was always the large companies populated by scores of engineers where I felt my skills were underutilized and where I was often ignored and could get little cooperation from the powers that be. At many of these places it seemed to me as if no one was driving the bus—or perhaps if someone was driving, I wasn't privy to any of the information that would have allowed me to see that someone was driving.

What to do about it? There has to be some education of managers and engineers about using the tech writer in the design process. I have always viewed my role as employed by the company but an advocate of the user. Too often the mindset of others in the company is that the user is the enemy (and thus the tech writer is too).

Analysis

The passion that these discussants have for the field of technical communication is obvious even if they sometimes are cast in roles as tragic heroes, working for the user in unappreciative environments. In this section, we examine the insights offered in the listserv discussion.

We focus specifically on the areas of disjuncture between the fantasy of being a technical communicator and the reality of that position. We address the changes the participants would like to see in their jobs and in the profession as a whole, and we investigate why so many of the participants have experienced burnout, causing them to change jobs or professions. To situate the experiences and concerns of the listserv participants within a larger framework, we draw on the literature in technical communication and organizational management.

Expectations and adaptations. Every one of the listserv participants expressed satisfaction with the education their degree program provided. However, most said that the training they received wasn't sufficient for the roles they played as technical communicators. Most technical communication degree programs probably do a good job of training their graduates to tackle communication tasks. Most notably, these tasks include writing, but potentially they also include tasks such as interface design and visual rhetoric in both paper and digital media. Yet there seems to be a range of additional skills that are necessary for professional success that are not taught (or at least were not taught to these students).

The listserv participants express the wish that they had been trained in interviewing methods (perhaps something more akin to ethnographic methods), business culture and fundamentals, marketing, planning, scheduling, and budgeting. As Alan stated,

> I was prepared to be a good technical writer. I was not prepared to deal with the environment of corporate America. I was not prepared to be a self-managed department responsible for estimating multiple, simultaneous project schedules with already-set deadlines.

There are other areas mentioned that are less easy to teach, like dealing with difficult SMEs and politicking, but the participants also have an appreciation that technical communication programs have to focus on the basics. Whether these additional topics are added to technical communication curricula in the future, it is important to know that there is an appreciable gap between what our students learn in our programs and what skills they apparently need on the job.

Another set of comments from the participants suggests that job expectations were shaped in specific ways by classroom and internship experiences, and that they experienced shock when reaching the work world. In the classroom, writing was highly valued and considered the focus of their roles as technical communicators. Even in their internships, they felt valued and were given opportunities to use several of the skills they had honed in the classroom. Within both of these learning environments, the students developed identities as critical thinkers, researchers, and writers, and their efforts to play out these roles were rewarded.

These classroom and internship opportunities, while providing valuable learning experiences for the students in the program, may have also prevented them from developing what one participant terms "a fair assessment of the real world." In the classroom, information is directly available or is laid out in an accessible repository (texts, the Web, and libraries). In an internship, projects are often well defined, and interns are sheltered from difficult personalities in the organization.

As the listserv participants experienced, workplace environments may look very different from the models represented in the classroom or even experienced through internships. There was likely to be much less variety in assignments and much less focus on writing as a percentage of the overall job. This emphasis on interpersonal information gathering, cajoling, and politicking seemed to be an unwelcome surprise for some of the participants. As Jill wrote,

> One of the things I liked about writing before I entered the profession was that it's a pretty solitary endeavor: you collect the information with as little ado as possible and then go home and write—or so I thought. Not so. The politics and coaxing information from people, etc. got old fast.

Individuals who are drawn to the profession out of a love of language and textual issues are not necessarily good at the aggressive data gathering that is often required.

Participants also seemed to have difficulty finding a well-defined role in their companies. While their roles as graduate students of technical communication were very well defined, becoming a professional technical communicator is not like becoming a dentist—where presumably you hang out a shingle and people with toothaches eagerly seek you out.

Some of the participants entered the workplace with the assumption that their value as technical communicators would not only be understood by others in the organization but also appreciated. In their graduate program they were trained to work collaboratively, but they were not prepared to have to convince others of their role and value. Consequently, the participants felt misunderstood, undervalued, and mistreated. There didn't seem to be an easy way to communicate about the mutual benefits of working together.

Roles and metaphors in technical communication. There have been a multitude of articles written in the last 10 years whose titles begin with "The role of the technical communicator in ..."—many more articles have "New roles" or "Changing roles" in the title. The Association of Teachers of Technical Writing call for proposals for their 2002 conference stated, "Today, identity formation continues to be a concern for many in the field." Among the questions posed by the call for proposals were the following: "Who are we? How are we perceived? Can our key role be acknowledged? Can we hope to attain the stature and the recognition we need to achieve our professional goals without such definition and identity?" (Association of Teachers of Technical Writing 2001).

This attention to identity and activity revision in technical communication derives from many converging factors: the rapid pace of technological change that continually presents technical communicators with new tools and new capacities; a changing technical economy where often information is the key component of the technology; and the maturing of the profession of technical communication and scholarship devoted to its study.

At the same time, our profession's identity formation results from the "immaturity" of the profession itself—as evidenced by how much of our lives we spend explaining to people what a technical communicator is. This convergence has spawned continuous reexamination of the roles that technical communicators do play, might play, and should/shouldn't play. Partly we have been trying to define the roles we play, and partly we have been searching for metaphors to represent our part in the technology enterprise.

At the beginning of this article, we heard the participants describe the many roles they found themselves playing. Most of them played the role of writer less than they imagined. Instead, they found themselves being programmers, educators, managers, designers, interrogators, investigators, planners, testers, and advocates for the user. Some of these roles were written into their job descriptions and others were extra roles they had to perform to get their jobs done. These roles aren't necessarily bad, but taking them on can be a bit like accepting a job that you think involves juggling tennis balls, and instead finding yourself juggling a tennis ball, a bowling pin, an ostrich egg, and occasionally a running chainsaw.

Other roles articulated by the group are a function of the lack of understanding employers and coworkers often have about technical communicators—the mismatch between how technical communicators think they can contribute and how others in the company think technical communicators can contribute. Perhaps the issue does come down to marketing and education.

As is often a theme in articles such as this one, it appears that technical communicators need to do a better job articulating their value and abilities to their employers. And as a profession, we need to better articulate our value and abilities to the industries that employ us. There are obstacles, certainly. Engineers and programmers are trained to think about the world differently than are technical communicators. As Winsor (1990) argues, knowledge-making for engineers often depends on physical reality and mastery of physical objects. Thus, it may be difficult for engineers to accept "textual mediation of knowledge." The engineering discipline's emphasis on production and machines can also lead engineers to "devalue texts" (58–59).

But as the engineering profession becomes more grounded in abstract, cyber-representations of physical objects, opportunities should present themselves for us to make connections as a profession with engineering. As they move farther away from their comfortably defined territory and we try to move toward a better-defined identity, perhaps we can coexist in new ways. Of course, such a shift would require a concerted effort by the profession—it won't be accomplished by thousands of individual technical communicators toiling in isolation from one another.

Also, part of our field's ongoing efforts to define our identity is the challenge of coming up with a good name for what we do. Boykin and Buonanno (1994) explore several different potential titles for technical communicators that might better reflect the value we contribute to companies: information developer, functional analyst, interface professional, project manager. They reject each of these because they don't quite get at what a technical communicator does, or because they emphasize production over synthesis, are too vague, or are too reminiscent of bridge metaphors.

Project manager is rejected because most projects already have managers, so to reflect the many activities of technical communicators, they suggest *information manager*. They find this term preferable because it encompasses all the positive attributes and responsibilities of the previous titles. It also encourages technical communicators to think of their role in the company broadly, potentially leading to further job opportunities:

> A technical writer may have joined an organization for a specific project, then move on upon the project's completion. Communicators who are on their toes and who demonstrate a wide range of abilities have greater opportunities to forge a new position. (Boykin and Buonanno, 542)

Another way to think about the position of technical communicators within organizations is, as Teresa M. Harrison and Mary Beth Debs (1988) suggest, from a systems approach. Such an approach, stemming from organizational theory, investigates the role technical communicators play within an organization composed of independent parts.

Through the communication activities of information acquisition, document reviews, and editing processes, technical communicators "engage in tasks that permit them to function as boundary spanners" (5). As boundary spanners, technical communicators are in a position that requires them to have a strong sense of a company's voice. Among other tasks, they speak with this voice to gain, interpret, and disseminate information from one group within an organization to another group within an organization. Yet, whereas this approach makes sense on a theoretical level, *boundary spanner* isn't exactly suitable for a business card.

Rather than focus on appropriate names for people who work in our profession, Janice Redish (1995) suggests another type of communication strategy. She advocates the quantification of the value added by technical communicators, asking how technical communicators impact the company's return on investment. She outlines approaches to demonstrate for managers how each of the multiple roles technical communicators play within a company directly impact the bottom line.

This practice, she says, is learning to speak the manager's language. From this perspective, the value of the technical communicator is reified for management and technical communicators could be called *squid-giggers* for all management cares, as long as they see the dollar signs.

But human business people do not act rationally 100% of the time (if they did there would be no *Dilbert* cartoons). Being valued within a particular company's culture seldom comes down only to dollars. Thus, the problem of being understood as valuable within the company in subjective terms remains critical.

We must note here that there is a duality to this discussion. On one hand, we are talking about professionals overwhelmed by the multiple roles they are asked to play, and on the other hand, we are talking about how to take advantage of and develop the roles technical communicators play to cement or expand employment opportunities. This duality seemingly derives from the lack of value most employers have for technical communicators—as noted in the listserv discussion.

There is instability in the role that technical communicators inhabit in the corporate world that is coupled with economic instability. In the tail of the early '90s recession, many technical communicators with large and stable companies found themselves downsized, and if they were lucky, then hired back as contractors without benefits. Likewise, in the dot-com bust, technical communicators were often among the first to be let go. The issue of well-defined roles cannot be dismissed as a trivial concern; the problem has real consequences for professionals in our field.

The participants on the listserv actually seemed to enjoy the expansion of their roles into new areas of responsibility, new technologies, or new challenges, as long as they were supported in those efforts and as long as those expansions didn't involve extra work arising from unreasonable obstacles that were placed in the way of doing their job. The listserv discussants came to understand that the role of technical communicator involves many tasks beyond sitting down to write or edit text (that is, programmer, educator, manager, designer, interrogator, investigator, planner, tester, and advocate for the user).

Perhaps many of these same burdens fall on the other employees of the company too, but the role of communication expert places these participants at certain cultural nexuses: they are in charge of representing the company and the product symbolically to the customer; they help construct the internal and external identity of the company. Their role permeates the company in a more complex fashion than just writing a book to put in the box. Technical communicators are *integrators* for projects and for companies. They must be conversant in the languages of every department and must integrate information, requirements, and constraints.

The role conflict the discussants encounter results from their failure (more likely the failure of the profession as a whole) and their managers' failure to arrive at an understanding of what technical communicators (can) do for their company and a failure to articulate a clear role that the technical communicators should play. Consequently, technical communicators operate to a large extent in management's blind spot, which can be a dangerous place.

Burnout within the technical communication industry. Burnout is in many ways the human equivalent of the learned helplessness described by Martin Seligman (1972). In his studies, test animals gave up on escaping from painful shocks and just lay down to suffer through their pain. Likewise, people experiencing job burnout are faced with daunting, insoluble problems that are highly stressful both physically and psychologically. As a result, workers may change professions (as several listserv participants did), or change employers (as several listserv participants did), or if they don't have the flexibility, mobility, or available options, the worker may stay in the troublesome work environment, as helpless as Seligman's animals.

According to the *Occupational Outlook Handbook*, "deadlines [and] erratic work hours," a common element among technical editing or writing jobs, "may cause stress, fatigue, or burnout" (2002). Results of an ongoing poll at the TECHWR-L Web site, a site devoted to current trends in technical communication, confirm the role burnout plays in our field. More than half of 189 professionals cited occasional burnout, almost a quarter cited frequent burnout, and a few reported "always" feeling job burnout (TECHWR-L 2003). The listerv discussants repeatedly mentioned job burnout as well.

Burnout is specifically defined as "overwhelming exhaustion, and feelings of frustration, anger and cynicism" (Maslach and Goldberg 1998, 63) that occur as a result of stress related to a job or activity. Cordes and Dougherty (1993) identify "role conflict, role ambiguity, and role overload" (623) as key components of burnout, and our discussion with the listserv participants revealed all three.

Role conflict—"when expectations from one's job conflict with one's values or personal beliefs"—appeared to have contributed to some of the feelings of burnout expressed by the participants in our discussion list. They were trained to approach writing tasks rhetorically and see communication as a vital part of product development and support, and they developed a certain image of the role they should play within an organization. When they were treated differently, perhaps as a "glorified secretary" or agent of the "grammar police," the clashing expectations added stress to their work.

Role ambiguity is another key factor for technical communicators. Our participants described working in environments where their work and their potential contribution were poorly understood by their managers and coworkers. This lack of awareness directly affects technical communicators' abilities to do their jobs, especially when coworkers do not recognize or value the proper steps, resources, or time, needed for the documentation process.

At the same time, technical communicators can easily experience role overload from the many hats they wear in an organization: programmer, educator, manager, designer, interrogator, investigator, planner, tester, and user advocate. As stated earlier, the technical communicator is in a unique position at the nexus of information and identity within an organization. But perhaps this position is not a comfortable one: demanding, poorly understood—not what they thought they were getting into when they started. As a result of role conflict, ambiguity, and overload, the listserv discussants reported experiencing a sense of worthlessness to the organization. As one of the participants stated, "I began to feel really burned out when I realized what little difference I was making for the companies I worked with, the reader, and society in general." Another listserv participant mirrored this sentiment, claiming: "when the time came that I felt I was no longer making a difference, I began to feel the burn."

In organizations where documentation is viewed as an afterthought and tangible products are considered much more valuable than the write-ups of these products, it may be difficult for technical communicators to feel and see rewards for their efforts. When they reach this point, technical communicators are often propelled toward making a major change. This change may take place on a local level and prompt the technical communicator to renegotiate job expectations and clarify roles within the work environment. The change may also take place on a global scale, involving revised career goals.

Technical communicators may be particularly vulnerable to burnout because of the types of people and personalities who are attracted to the profession, and because of the service mindset that is a part of how we understand what technical communicators do. Since the mid-1970s when the term was first coined, many of the studies of job burnout have focused on professions like teaching, nursing, and counseling—service jobs where there are large demands. As Christina Maslach and Julie Goldberg (1998) write,

> The norms of these types of caregiving, teaching, and service occupations are clear, if not always stated explicitly: to be selfless and put others' needs first; to work long hours and do whatever it takes to help a client, or patient, or student; to go the extra mile and give one's all (63).

While technical communicators are not technically teachers, nurses, or counselors, it certainly feels that way much of the time. And even though we don't aspire to the norms cited in the quotation, the role of technical communicator is often to conform to such expectations. In many circumstances, we must be selfless and put others' needs first because we are made to believe that the jobs that the engineers and programmers do are much more important than what we do.

We must work long hours and do whatever it takes to help the client, but we have two clients: the actual person who will use the product and the SMEs whose products we document. One client we work for but never see. The other client we work to please but often relate to in a dysfunctional way. We go the extra mile because we believe in the value of what we do, but we wonder whether less heroism would be required of us if production tasks were just structured differently.

A recent study of job burnout suggests that traditional approaches to addressing burnout have been mistaken in their focus on the individuals experiencing the condition. Traditional approaches have suggested changes in work patterns (work less), development of coping skills, development of support networks, development of relaxation strategies (meditation, massages, hot baths, and so forth), improvements in health (diet and exercise), or improvements in self knowledge (learning how you need to change to deal with stress). It was assumed that the individual needed to change, because stress was a natural if not essential part of work (Maslach and Goldberg, 67–68).

A different approach advocated by this study, however, is to understand the role that the work environment plays—and more importantly, to understand how the relationship between the individual and workplace contribute to burnout. Ideally, a job should promote energy and not exhaustion, involvement and not detachment, and feelings of effectiveness instead of diminished accomplishment (Maslach and Goldberg, 66).

One part of that goal is helping employees understand the burnout risks associated with their jobs and the ways that their decisions as workers can contribute to those risks. For example, deciding one week to take on more projects than you can handle can quickly develop into a pattern of continually taking on more work than you can handle. This increased workload can easily put an employee on the path to burnout (Maslach and Goldberg, 70).

As Alan suggests, burnout is a potential "issue for any technical writer who has a passion for what he/she does." This passion is also connected with Mara's attribution of burnout as "related to boilerplating and unappreciative coworkers." Ours is a profession that attracts people with a strong belief in the calling of clear communication, and our degree programs do a good job of reinforcing those inclinations.

We all collect parables (horrifying or humorous) of the disaster or foolishness that resulted from miscommunication or malapropism. These parables are more than the geeky inside jokes of language nerds; they are the mythology that supports our belief in our profession.

When technical communication students graduate and enter the working world, they are sometimes met with repetitious monotonous work like proofreading package inserts printed in six-point type or stringing boilerplate together, and sometimes they find themselves with numerous challenges for which they are not prepared. They are driven by the will to do the right thing for the engineers and the users, but this drive will only triumph over experience for so long before they move on to something else, wondering where the failure occurred.

Conclusion

Failure, however, is not where we choose to end this piece, and it was not the loudest message resonating from the listserv. Although the participants pointed to numerous challenges they faced when entering the technical communication profession and cited complex work scenarios they continue to face, the key to overcoming all these challenges resides in something all technical communicators possess: clear communication skills.

To bridge the gap between academic preparation and workplace realities, it is necessary to continue to maintain open lines of communication with graduates and practitioners working in industry. As Louise Rehling (1998) observes, we should seek opportunities to "exchange expertise" with practitioners in industry so that we may learn from the workplace, and in turn, educate it too.

Our field must also continue efforts to communicate the roles we can and cannot play, and the value we can bring to work groups and to documentation processes and products. Communication skills can also help us to recognize and diffuse the on-the-job circumstances that so often result in burnout. In particular, we share a number of suggestions, inspired by the experiences of the listserv participants.

Suggestion 1. Include an emphasis on general business concepts in technical communication curricula. Because technical communication students will not necessarily have a background that includes business classes, a course that provides an overview of basic business principles would be useful. Such a course may include an introduction to product life cycles, project management, and functions of engineering, information systems, and marketing departments. If all technical communication students (we realize that some students may already have this training or equivalent experience) possessed an elementary understanding of the ways in which businesses operate, their learning curve on the job would not be as steep.

Suggestion 2. Teach students to market themselves as valuable contributors to an organization. To prepare students to work with individuals who may not understand or appreciate the value a technical communicator brings to an organization, as educators we should spend time discussing with students the value of our profession.

Carliner's summary of the value communicators add (1998), including specific ways in which they affect an organization's bottom line, contains information that attests to the value of technical communicators. Case studies from various industries, such as those of Fisher and Sless (1990), Dunlap (1992), Blackwell (1995), and Pieratti (1995), provide quantitative and qualitative accounts of the ways in which technical communicators helped to improve customer relations, reduce costs, and improve overall organization within different companies. Henry's study of the ways in which a technical communicator added value to a project team (1998) also provides evidence of the expertise communicators possess and the ways in which others within an organization benefit from this expertise.

Exposing students to these studies grants them knowledge of how they can contribute to an organization, and it also arms them with the confidence to market themselves to others who may question their capabilities.

Suggestion 3. Give students practice obtaining difficult information. Several of the participants discussed the difficulty they faced obtaining information that they needed to do their jobs. Whether the information-gathering took the form of interviewing an engineer about a product or picking the brain of a subject matter expert, the technical communicators learned the hard way that information is not always readily handed over. The experience surprised the participants, in part, because it contrasted with class projects and internships where coworkers willingly gave them the data they asked for.

We suggest that educators avoid assignments that are so "neat" in specification of the task they lack a challenge for the student. Following Grice's argument (1997) that assignments that are too specific and detailed deprive students of "acquiring skills and a flexible attitude toward their work they will need later in their careers," we should consider presenting students with tasks that are complex and contingent on cooperation of others (219).

Although we may worry that such "messy" tasks might land students in awkward and frustrating situations, providing them with this practice will make future experiences less shocking and perhaps more easily managed. In addition, if students are encouraged to share these experiences in the classroom and then work collaboratively toward solutions, we can help them learn problem-solving and negotiation strategies that they can later apply in the workplace.

Suggestion 4. Appoint experienced mentors to newly hired technical communicators. To assist new technical communicators in their process of becoming acculturated to the organization and their particular job responsibilities, we see benefits to linking seasoned technical communicators with newcomers. A mentoring program, which already exists in several companies, gives newcomers the chance to ask questions, communicate concerns, and learn from a valuable resource—someone who has already been through the socialization process.

Suggestion 5. Involve technical communicators on project teams to emphasize the value they bring to company processes and final products. While organizations within some industries have already adopted this model (several within the software industry, for example), we should continue to educate industry about the benefits of including technical communicators in work groups.

If they are involved with projects from the beginning stages, technical communicators are more likely to demonstrate their skills in planning and conceptualizing documents and communication processes. Working with technical communicators during project conceptualization meetings can help engineers, technology experts, and team members from other areas of expertise realize that technical communicators do more than just "fix the language" of the final product.

Suggestion 6. Support professional growth of technical communicators and demonstrate that they are valued by the organization. This suggestion, targeted toward technical communication managers, comes directly from Alan Sloan's listserv response regarding what managers can do to help their employees prevent or alleviate burnout.

Acknowledgments

The authors would like to graciously thank Alan Sloan, Tom Kolano, Jill Ann Mortell, Monica Mitchell, Mara Cohen Ioannides, Bill Pollak, and Margo Parent for their assistance with this article.

References

Association of Teachers of Technical Writing. 2001. Call for proposals. http://www.attw.org

Blackwell, C. 1995. A good installation guide increases user satisfaction and reduces support costs. *Technical Communication* 42:56–60.

Boykin, Carolyn, and Elizabeth Buonanno. 1994. Alternative names for the roles of the communicator. In Theory and the profession: Evolving roles of the communicator, ed. Charles E. Beck. *Technical Communication* 41:539–543.

Carliner, Saul. 1998. Business objectives: A key tool for demonstrating the value of technical communication projects. *Technical Communication* 45:380–384.

Cordes, Cynthia L., and Thomas W. Dougherty. 1993. A review and an integration of research on job burnout. *The Academy of Management Review* 18:621–657.

Dunlap, Johnny. 1992. Customer satisfaction: The documentation challenge. *Proceedings of the 39th International Technical Communication Conference*, 700–703, Arlington, VA: Society for Technical Communication.

Fisher, Phil, and David Sless. 1990. Information design methods and productivity in the insurance industry. *Information Design Journal* 6:103–129.

Grice, Roger A. 1997. Professional roles: Technical writer. In *Foundations for Teaching Technical Communication: Theory, Practice, and Program Design*, ed. Katherine Staples and Cezar Ornatowski, 209–220. Greenwich, CT: Ablex Publishing Corporation.

Harrison, Teresa M., and Mary Beth Debs. 1988. Conceptualizing the organizational role of technical communicators: A systems approach. *Journal of Business and Technical Communication* 2:5–21.

Henry, Jim. 1998. Documenting contributory expertise: The value added by technical communicators in collaborative writing situations. *Technical Communication* 45:207–220.

Maslach, Christina, and Julie Goldberg. 1998. Prevention of burnout: New perspectives. *Applied and Preventative Psychology* 7:63–74.

Occupational Outlook Handbook. 2002. http://www.bls.gov/oco/ocos089.htm

Pieratti, Denise D. 1995. How the process and organization can help or hinder adding value. *Technical Communication* 42:61–68.

Redish, Janice. 1995. Adding value as a professional technical communicator. *Technical Communication* 42:26–39.

Rehling, Louise. 1998. Exchanging expertise: Learning from the workplace and educating it, too. *Journal of Technical Writing and Communication* 28:385–393.

Seligman, M. E. P. 1972. Learned helplessness. *Annual Review of Medicine* 23:407–412.

TECHWR-L. 2003. http://www.raycomm.com/techwhirl/polls/techwhirlpollresults.php3?pollID=51

Winsor, Dorothy. 1990. Engineering writing/writing engineering. *College Composition and Communication* 41(1):58–70.

Commentary

If you are a technical communication student, you probably found Wilson and Ford's study interesting—possibly even sobering—because it provides a glimpse of what you may find when you complete your degree and get your first job. If you are already in the workplace, perhaps you found yourself nodding in agreement at some points and shaking your head vigorously in disagreement at others. But whether you are preparing for a career in technical communication or have already begun one, it is unlikely that you found this article dry or boring.

One of the hallmarks of good qualitative research reports is the rich grounding in the area of interest they usually provide, regardless of their subject. The fact that Wilson and Ford's article addresses such central topics for technical communicators—and those who teach and mentor them—makes its evocation of the "Big Chill" even more compelling to technical communicators.

There is no doubt that the focus group transcript contained in the article is largely responsible for the sense of reality that pervades it. Although studies based on focus group research typically contain extended quotations and paraphrases from the discussions, they rarely reproduce the entire conversation because it would seldom command the degree of interest for the audience that this dialog holds. As a result, peer reviewers would not typically recommend manuscripts for publication when they reproduce the discussion, and journal editors would seldom be able to justify the space devoted to such transcripts.

Purpose and Audience

In this article, Wilson and Ford explore the research question stated in the first paragraph:

> As researchers, we were particularly interested in knowing how the educational training of [individuals who have been in the technical communication field for over a decade] influenced their acclimation to the work world. We were also interested in understanding the roles they played within their organizations, as well as comprehending the ways in which they interacted with coworkers, and identifying the triggers for feelings of burnout and disillusionment that many of the subjects experienced.

As with the other articles reprinted from *Technical Communication*, we can assume that this study is directed toward the dual audience of the Society for Technical Communication: practitioners and academics. Young practitioners will appreciate the article's candid and unapologetic treatment of the problems that people like them often face in our field. More experienced communicators in business and industry, especially managers and project leads, will find the article helpful in career counseling and helping colleagues develop professionally. And teachers of technical communication will value the insights the report can provide them in evaluating courses and programs.

Organization

As mentioned before, this article is unusual because of the extended transcript of the focus group discussion that it contains. Otherwise, though, its organization is typical of this type of report.

- The Introduction states the research question, describes the methodology for the study, and lists the names, positions, and backgrounds of the seven focus group participants.
- The Listserv Discussion provides an edited transcript of four questions with extended answers from the participants. This section accounts for more than a third of the article.
- The Analysis section categorizes the discussion under three major topics: Expectations and adaptations, roles and metaphors in the field, and burnout.
- The Conclusion offers six suggestions "To bridge the gap between academic preparation and workplace realities ...."
- The Acknowledgments recognize the focus group participants.
- Although the article doesn't contain a literature review, the References section provides citations of the secondary works mentioned in the article. Virtually all of these are discussed in the Analysis and Conclusion sections and are used to place the insights from the focus group and the recommendations based on the analysis in the context of the insights of other experts in the field.

Qualitative Methodology

The Virtual Focus Group. The methodology of this study is the focus group, but the researchers have used listserv technology to conduct a virtual focus group. Instead of gathering the participants in a room with a facilitator and recorder for several hours, Wilson and Ford asked the list members a series of questions by e-mail and then collected the responses to the questions and to other list members' responses. The result is a good approximation of the traditional focus group technique.

The advantage of this virtual focus group is that it did not require the researchers to gather the participants in one location. Because the focus group members were geographically dispersed, this advantage was significant—it allowed the researchers to select participants entirely on the basis of characteristics that they had defined for inclusion in the study rather than on proximity to a meeting place. Of course, this problem could have been overcome by paying the travel expenses of participants to meet in a central location, but such a solution would be more costly than practicable for almost all studies in our field.

The use of listserv technology has another important advantage. Because all the responses are contained in e-mail messages, there is no need to record and transcribe them. As a result, the researchers saved time as well as money.

Of course there are some disadvantages that to some extent balance the advantages of the virtual focus group. When participants are physically collocated for a focus group discussion, the result is very spontaneous, with group members frequently piggybacking off of the ideas of others. The result is that the total effect of the discussion is often much more than the sum of the individual participants' contributions. This spontaneity is lost at least in part when the group members must read the thoughts of others—perhaps days or weeks later—and then reply.

Another disadvantage is that while the timeframe of the traditional focus group is limited to no more than two or three hours, the virtual focus group must take place over at least several days. In this case, the discussion lasted five months. As a result, the time commitment of participants was significantly increased, and the time saved by not having to record and transcribe the discussion was certainly lost. Similarly, though a skilled facilitator can generally retain the participants' interest in the discussion fairly easily for a few hours, it is much more difficult to do so when the discussion lasts for an extended period.

The Participants. The seven participants in the virtual focus group all graduated from the same master's program in technical communication in 1991 and 1992. They were recruited through a message on the program's alumni listserv, and though the authors do not say so explicitly, it may be that these seven people were the only alumni who agreed to participate (Wilson and Ford note that the focus group consisted of "the seven volunteers"). We do not have a sense of how large a percentage of the alumni from 1991–1992 these seven represent, but it is likely that they constitute a significant proportion as most master's programs in technical communication at that time were modest in size.

Based on what we know about quantitative research and surveys, it may seem that the sampling method for this study was problematic because the participants are not likely to be representative of the total population of technical communicators. They all have similar educational backgrounds and, in fact, are all alumni of a single master's program. All have roughly the same amount of professional experience, and they are probably less geographically dispersed than a random sampling of technical communicators from around the United States, much less around the world.

It is important to remember, though, that qualitative studies are different from quantitative research in that they do not use inferential statistics. In other words, there is no attempt to generalize about a larger population based on the results seen in the sample population.

In fact, Wilson and Ford make some interesting observations about the representativeness of their sample.

> One could question how representative this conversation and its occasional criticisms are of technical writers as a whole. It was not our intention to do a scientific survey of attitudes across the profession—many studies of that type have been done. Our goal was to allow the voices of professionals to relate their experiences, and one must trade off breadth of coverage when seeking that kind of depth.
>
> From a similar starting point, however, the seven participants in the listserv have been involved in many different professional situations (for example, engineering firms; dot-com startups; education; consulting; software, hardware, and biomedical firms) in companies of various sizes, and are able to discuss a wide range issues pertinent to our field. And as graduates of a master's-level technical communication program, arguably they are representative of professionals who enter practice in the field with the most preparation and training, representative of the types of students current degree programs are training.
>
> As such, they may be somewhat unrepresentative because they were better prepared to deal with the problems they discuss than many who enter the field, but still, there is much to learn from their comments.

The authors are completely right here. The sample population is not representative, and the number of participants is well below the threshold for a survey. Although including alumni of multiple master's programs might have provided more diversity of opinion and experience, the resulting focus group would have lost its quality reminiscent of the 1983 film *The Big Chill*, in which seven college friends gather for the funeral of one of their colleagues a decade after graduation.

Moreover, the size of the sample population is not relevant in a qualitative study. For example, the "discount" approach to usability testing that has been the norm since the 1990s requires only three to five testers rather than the 30 to 50 that were commonly used in the more "scientific" usability studies before that time (Barnum 2002, 10). In qualitative research, breadth of the study population and the probability that the results are representative of the general population are traded off for depth of the data collected and richness of detail available for analysis.

We will address the representativeness of the sample again when we talk about the study's credibility.

Transcript

As Wilson and Ford report in the article's Methods section, the focus group discussion was conducted between mid-October 1999 and mid-March 2000. The researchers began by asking each participant to contribute a narrative recounting their experiences since completing the program. Once all the narratives had been submitted, they began posting questions, and the discussion on that question continued until "traffic on the list died down," though Wilson and Ford did occasionally prompt the participants to continue if list traffic abated before they sensed that the question had been discussed in full.

Although the first two questions (expectations on entering the profession and the most important lessons learned on the job) were planned in advance, the two final questions (burnout and roles played in the workplace) grew out of the discussion. This flexibility is typical of focus groups generally, though the extent of that flexibility is not practical for face-to-face focus groups conducted in a few hours. In this case, the researchers had lots of time to confer about the next question they wished to ask based on the direction of the discussion to that point.

Wilson and Ford note that they "did not attempt to moderate or contribute" to the discussion (though they did contribute professional narratives at the beginning of the process "so that the participants would be more comfortable with those who were asking the questions"). The transcript does not include the narratives. The authors also observe that they have edited the transcript to shorten it and have also deleted off-topic discussion.

Because we know that the transcript is not complete, our analysis of the content must necessarily be tentative, but it is interesting to note the following characteristics.

- For Question 1 (expectations entering the profession), the transcript contains eight comments and responses representing six of the seven participants.
- For Question 2 (lessons learned on the job or about the profession), there are seven comments and responses from five of the seven focus group members.
- For Question 3 (burnout), the transcript provides four responses and comments representing four of the seven participants.
- For Question 4 (roles in the workplace), the transcript reports five comments and responses from four of the seven focus group members.

The number of contributions by the seven participants varies widely. Bill, for example, answered only the first question, and Monica, only the first two, while Margo contributed six responses, and Mara, seven. How much of this variation is the result of the authors eliminating repetitive responses and how much it represents the differing patterns of responsiveness of the focus group members cannot be known. The distribution of responsiveness among these participants is not surprising. After all, some people are talkative, while others are reticent. In a face-to-face focus group, however, the facilitator might well have encouraged Bill and Monica to participate more.

Analysis

Coding and Categorizing. The coding and categorizing of data produced in a qualitative research project is done "off stage." The reader does not see the researchers doing this work, just the results. On the basis of the analysis provided in the article, however, we know that the researchers used at least the following codes to classify the data from the transcript:

- Expectations about the workplace, which might be coded EW
- Expectations about the profession, which might be coded EP

- Positive roles technical communicators assume in the workplace, which might be coded PR
- Negative roles technical communicators play on the job, which might be coded NR
- Metaphors and names for what technical communicators do, which might be coded M/N
- Burnout, which might be coded B

Although we do not know for certain whether these codes were predefined or open, the authors may have determined them very early in their research project because of the abundance of articles in the literature, cited in their Analysis and Conclusion sections, that reflect these concerns within the profession. In fact, it is possible that the authors combined the coding and categorizing stages of analysis for this project.

The five codes might be collapsed into the following three categories:

- Expectations
- Roles, metaphors, and names
- Burnout

Indeed, these categories correspond almost perfectly to the three subsection titles within the Analysis section of the article.

Exercise 9.1

Obtain the text of the focus group discussion from the book Web site (www.ghayhoe.com/tcresearch/focus.doc). Open the document in Microsoft Word, read it carefully, and break it into chunks consisting of phrases or sentences. Then apply open codes using the Word indexing tool. Revise your index to consolidate codes, and then compile the index. ■

Exercise 9.2

Using your response to Exercise 9.1, can you identify other categories that Wilson and Ford might have explored in analyzing the responses of focus group participants? ■

Detecting Patterns or Modeling. In the case of this study, the need to discern patterns or construct models from the data that has been coded and categorized is almost unnecessary. The dataset collected by the researchers (slightly more than 5300 words in the edited transcript published in the article) was limited by comparison with the output from a typical two-hour focus-group session. Moreover, the direction of the discussion, at least in the latter two questions, was determined by the research team in response to concerns revealed by the participants in their answers to the first two questions. In short, the patterns in the data are identical to the categories.

Standards of Rigor and the Conclusions of the Study

Although derived from the focus group discussion, the conclusions that Wilson and Ford present in the form of six suggestions to faculty of academic programs and managers of technical communication practitioners echo recommendations by authors of other articles and conference presentations.

1. Include an emphasis on general business concepts in technical communication curricula.
2. Teach students to market themselves as valuable contributors to an organization.
3. Give students practice obtaining difficult information.
4. Appoint experienced mentors to newly hired technical communicators.
5. Involve technical communicators on project teams to emphasize the value they bring to company processes and final products.
6. Support professional growth of technical communicators and demonstrate that they are valued by the organization.

We need to examine these conclusions in terms of the standards of rigor for qualitative research.

Credibility. In Chapter 5 we noted that in a quantitative study we assess the internal validity by essentially asking whether the researchers have measured the concept they wanted to study, whereas in a qualitative study, we look more to the credibility of the data. Did the participants truly represent the population or phenomenon of interest and are their behavior and comments typical of that population? The example we cited was a study of how help-desk personnel use a product's technical documentation based on interviews of help-desk supervisors. We concluded that help-desk supervisors might not be credible sources of information about how help-desk employees use documentation.

The question then is whether the seven master's program alumni who participated in Wilson and Ford's focus group are credible representatives of the population of interest. As the researchers have described their population of interest in very narrow terms—graduates of a specific program from 1991–1992—we can conclude that these participants give the study credibility.

It is important to note that when we talk about credibility of the population of a qualitative study, we do not mean that this is a representative sample of a much broader population. The researchers do not intend to generalize about all master's graduates 10 years out of school. Instead, they are conducting a discussion with a small group to explore experiences and problems they encountered. Furthermore, this study used the focus group methodology credibly to discover the participants' opinions, motives, and reactions, not to learn about their behavior.

Transferability. The requirement of external validity in quantitative research (whether the thing or phenomenon being measured in the test environment reflects what would be found in the real world) is comparable to the need for transferability in qualitative research (whether what we observe in the test environment reflects what would be found in the real world). Is the discussion that Wilson and Ford report authentic? Do their participants reflect concerns and experiences that could be discerned if we listened to discussions conducted with other groups of technical communicators with similar backgrounds? As we observed in Chapter 5, the burden here is on the reader of the study, but the fact that the literature abounds with examples that parallel Wilson and Ford's suggests that transferability is not a problem.

Dependability. Just as quantitative research must be reliable, qualitative research must be dependable. How confident can we be that the conclusions reached in the research project could be replicated if the study were conducted by different researchers? As we noted in Chapter 5, we make this determination based on the researchers' depth of engagement, the diversity of their perspectives and methods, and their staying grounded in the data.

Depth of Engagement. By depth of engagement, we mean that the more opportunities researchers give themselves to be exposed to the environment and to observe the data, the more dependable their findings will be. Using the focus group method, researchers would typically gain depth of engagement by conducting multiple discussions with multiple groups or by meeting with the same group on several occasions over a period of time. Because the single focus group on which Wilson and Ford base their findings was conducted over five months, however, this study meets the requirement of depth of engagement.

Diversity of Perspectives and Methods. Seeing data from multiple perspectives—for example, using multiple researchers or multiple data collection techniques—increases the rigor of a qualitative study. Because this research project was conducted by two people, it is characterized by at least a minimal degree of diversity of perspective. Diversity of methods is more problematic, however.

Using multiple methods such as field observations, interviews, and document analysis allows researchers to compare the data collected using the various techniques to determine whether it is internally consistent and whether the conclusions suggested by the data gathered from field observation, for example, can also be drawn from the data collected from the interviews and document analysis. Although Wilson and Ford arguably used two methods in their study (professional narratives that could be subjected to document analysis and focus group discussions), they do not summarize the professional narratives or make any explicit connections between information contained in those narratives and information elicited from participants in their responses to the four discussion questions.

Furthermore, the participants in the focus group are not diverse. As we have already noted, they are drawn from a very narrow population of interest, so we must ask whether there are other ways that this study achieved diversity of perspective.

Using another technique called member checking, researchers ask the participants to review their analysis of the data and the conclusions based on it—a kind of "sanity check," if you will. But there is no indication that Wilson and Ford have used member checking.

Yet another way of ensuring diversity of perspective is peer review—asking other researchers whether the conclusions reached in the study make sense based on their own experience. This type of peer review is typically performed during the study by asking colleagues to examine interview transcripts, field notes, or usability test results, as well as tentative conclusions based on that data. Again, there is no evidence in Wilson and Ford's article that they used this technique, though the manuscript itself was peer reviewed and approved by the reviewers prior to publication in *Technical Communication*.

So, diversity of perspectives and methods is a weakness in this study. However, the fact that what Wilson and Ford report is reinforced by so many other investigations of their topic suggests that this weakness is not significant.

Staying Grounded in the Data. The final determinant of dependability is staying grounded in the data. All conclusions and statements must be traceable back to directly observed data within the study. This is certainly not a problem with Wilson and Ford's study. All of the six suggestions in their Conclusion can be mapped to the focus group discussion.

Summary

In this chapter we have analyzed a sample article reporting the results of a qualitative research study using the concepts and methods presented in Chapter 5. Following the text of the exemplar article, we have examined its purpose and audience, its structure, the study's methodology, the transcript of the virtual focus group, the analytical methods that were likely used to interpret the data, the standards of rigor, and the credibility, transferability, and dependability of the study's conclusions. The conclusions resulting from the focus group discussion reported by Greg Wilson and Julie Dyke Ford meet these standards of rigor to an acceptable degree because the participants represent the population of interest (though that population is significantly more limited than is true in the typical qualitative study), because the discussion they report is authentic, and because the study's conclusions could very likely be duplicated by different researcher.

References

Barnum, C.M. 2002. *Usability Testing and Research*. New York: Longman.

Wilson, G., and J.D. Ford. 2003. The big chill: Seven technical communicators talk ten years after their master's program. *Technical Communication* 50:145–159.

Answer Key

Exercise 9.1

The answer to this exercise will be unique for each person who prepares it, so there is no key to this exercise.

Exercise 9.2

The answer to this exercise will be unique for each person who prepares it, so there is no key to this exercise.

10

Analyzing a Report on the Results of a Survey

Introduction

Chapter 6 explored surveys in terms of what they can measure, what kinds of questions they can ask, how questions can be formatted, and how they should be structured. In Chapter 6 we also discussed how to report results, looking especially at frequency distribution, response rate, and margin of error and confidence intervals. In this chapter, we examine in detail a journal article that reports the results of a survey. The chapter contains the full text of Sam Dragga's "'Is this ethical?' A survey of opinion on principles and practices of document design," which appeared in the August 1996 issue of *Technical Communication*, along with a commentary on the article.

Learning Objectives

After you have read this chapter, you should be able to

- Analyze an article reporting the results of a survey
- Apply the results of your analysis to preparing your own report of survey results

The Article's Context

Most technical communicators, both in academe and in industry, have degrees in fields outside this discipline—in English language or literature, rhetoric, composition studies, communications, or another field. As a result, the vast majority of people working in our field must apply general ethical concepts they have learned in philosophy or business courses to the specialized problems of technical communication in the workplace. In addition, few technical communication teachers have experience working in industry, so their familiarity with the ethical challenges of the workplace may be quite limited.

In reporting the results of a brief survey on document design ethics administered to samples of both practitioners and teachers, Dragga finds that our field lacks a guiding philosophy based on considered practice. Because the article is now more than 10 years old and because there seems to be a lot of turnover in the ranks of technical communicators—people join and then often leave the profession after only a few years—we should not automatically assume that the findings of Dragga's study are applicable to technical communicators today. Nevertheless, this article provides us with a lot of food for thought, and many teachers use this article in course modules on technical communication ethics because it provokes lots of questions and fascinating discussions.

"'Is This Ethical?' A Survey of Opinion on Principles and Practices of Document Design"*

Sam Dragga

Introduction

This article reports the results of a national survey of technical communicators and technical communication teachers regarding their perspectives on the ethics of various document design scenarios. But before you start reading this article, I would like you to answer the seven questions on this survey yourself (see Figure 10.1). In doing so, you will give yourself the opportunity later in this article to examine your answers relative to the findings of the national survey without being biased by the findings described here. You will also prepare yourself to read this article critically, thinking through with me the ethical issues raised by the rhetorical power of document design.

Rhetorical Power and Ethical Obligations

Before the computerization of verbal and visual communication—the days of pencils and typewriters—technical communicators were technical writers. The writer's only job was composing words. Graphic artists did the illustrations, and compositors and editors designed the pages. Today, more and more often, the technical writer is a technical communicator, choosing the typography and graphics as well as the words, designing the pages as well as checking the spelling. This ability to design information gives the technical communicator a new rhetorical power and imposes new ethical obligations on using that power.

This new rhetorical power, however, is also a source of peril for technical communicators because little research or guidance is available to identify the principles and practices that would lead to ethical document design. For example, *The Ethics of Human Communication* (Johannesen 1990), a widely cited book on the subject of ethics, dedicates two pages to the ethical dimension of "nonverbal communication" and asks a series of unanswered questions. The STC "Code for Communicators" (revised 1988) is also of little aid, with the exception of advising a communicator to "hold myself responsible for how well my audience understands my message." This advice (as we'll see later) comes close to a philosophy of ethical document design, but it is also buried as the fourth item in a bulleted list of seven professional guidelines, at least three of which have nothing to do with ethics.

Similarly, articles in the major journals of the field characterize ethics exclusively as a verbal issue (Bryan 1992; Buccholz 1989; Clark 1987; Radez 1980; Rubens 1981; Sachs 1980; Shimberg 1980; Walzer 1989; Wicclair and Farkas 1984). In a 1987 *Technical Communication* editorial, however, Girill mentions graphics

* This article was originally published in *Technical Communication* 43:255-265. References' and citations' styles have been changed to conform to *The Chicago Manual of Style*, 15th edition. Reprinted with the permission of the Society for Technical Communication.

Figure 10.1 Survey of ethical choices on document design.

1. A prospective employer asks job applicants for a one-page resume. In order to include a little more information on your one page, you slightly decrease the type size and the leading (i.e., the horizontal space between lines). Is this ethical?

1	2	3	4	5
Completely ethical	Mostly ethical	Ethics uncertain	Mostly unethical	Completely unethical

Please explain:

2. You are preparing an annual report for the members of the American Wildlife Association. Included in the report is a pie chart displaying how contributions to the association are used. Each piece of the pie is labeled and its percentage is displayed. In order to deemphasize the piece of the pie labeled "Administrative Costs," you color this piece green because cool colors make things look smaller. In order to emphasize the piece of the pie labeled "Wildlife Conservation Activities," you color this piece red because hot colors make things look bigger. Is this ethical?

1	2	3	4	5
Completely ethical	Mostly ethical	Ethics uncertain	Mostly unethical	Completely unethical

Please explain:

3. You have been asked to design materials that will be used to recruit new employees. You decide to include photographs of the company's employees and its facilities. Your company has no disabled employees. You ask one of the employees to sit in a wheelchair for one of the photographs. Is this ethical?

1	2	3	4	5
Completely ethical	Mostly ethical	Ethics uncertain	Mostly unethical	Completely unethical

Please explain:

4. You have been asked to evaluate a subordinate for possible promotion. In order to emphasize the employee's qualifications, you display these in a bulleted list. In order to de-emphasize the employee's deficiencies, you display these in a paragraph. Is this ethical?

1	2	3	4	5
Completely ethical	Mostly ethical	Ethics uncertain	Mostly unethical	Completely unethical

Please explain:

parenthetically: "Truthfulness requires that although we condense technical data, we should not misrepresent them to our audience (we can suppress the data points, e.g., but the curve should still have the same shape as before)" (178). Perica's 1972 *Technical Communication* article is also a curious exception. Writing prior to the computerization of communication technologies, Perica offers a brief list of guidelines regarding document design. He declares that airbrushing photographs to "highlight essentials" is ethical, but deleting "unsightly or unsafe items" is unethical. Using special typography, color, or glossy photographs is ethical unless important information is obscured. Double-spacing and using wide margins to make a publication look longer is ethical; single-spacing and using narrow margins to make a publication look shorter is also ethical.

While studies of technical communication ethics typically omit the subject of document design, research focusing on document design usually offers little

Figure 10.1 (continued) Survey of ethical choices on document design.

5. A major client of your company has issued a request for proposals. The maximum length is 25 pages. You have written your proposal and it is 21 pages. You worry that you may be at a disadvantage if your proposal seems short. In order to make your proposal appear longer, you slightly increase the type size and the leading (i.e., the horizontal space between lines). Is this ethical?

1	2	3	4	5
Completely ethical	Mostly ethical	Ethics uncertain	Mostly unethical	Completely unethical

Please explain:

6. You are preparing materials for potential investors, including a 5-year profile of your company's sales figures. Your sales have steadily decreased every year for five years. You design a line graph to display your sales figures. You clearly label each year and the corresponding annual sales. In order to deemphasize the decreasing sales, you reverse the chronology on the horizontal axis, from 1989, 1990, 1991, 1992, 1993 to 1993, 1992, 1991, 1990, 1989. This way the year with the lowest sales (1993) occurs first and the year with the highest sales (1989) occurs last. Thus the data line rises from left to right and gives the viewer a positive initial impression of your company. Is this ethical?

1	2	3	4	5
Completely ethical	Mostly ethical	Ethics uncertain	Mostly unethical	Completely unethical

Please explain:

7. You are designing materials for your company's newest product. Included is a detailed explanation of the product's limited warranty. In order to emphasize that the product carries a warranty, you display the word "Warranty" in a large size of type, in upper and lower case letters, making the word as visible and readable as possible. In order to deemphasize the details of the warranty, you display this information in smaller type and in all capital letters, making it more difficult to read and thus more likely to be skipped. Is this ethical?

1	2	3	4	5
Completely ethical	Mostly ethical	Ethics uncertain	Mostly unethical	Completely unethical

Please explain:

discussion of ethical issues or implications (Benson 1985; Dragga 1992; Felker and colleagues 1981; Kostelnick 1990; Murch 1985; Schriver 1989; White 1982). Two important exceptions are Edward Tufte's *The Visual Display of Quantitative Information* (1983) and Mark Monmonier's *How to Lie with Maps* (1991). Tufte's book, however, discusses only the ethics of graphical display, offering guidelines without evidence of their practical merit. Monmonier's perceptive book is equally restrictive, focusing exclusively on the ethics of mapping.

Are clear professional guidelines or a substantial body of research necessary to guide the ethical exercise of this new rhetorical power? Without a guiding philosophy, do technical communicators espouse ad hoc and erratic practices? Bryan (1992) believes that neither codes of conduct nor journal articles on ethics are effective motivators of ethical behavior because practicing professionals typically ignore guidelines and theoretical discussions, preferring books and magazines that identify specific strategies for success on the job. Walzer (1989) also has criticized the existing research and questioned the impact of codes of conduct on ethical behavior: "More important than even the best set of proscriptions is the complex moral sensibility that can only be honed by confronting and discussing difficult questions of ethics" (105).

A Survey of Opinion on Ethics

To initiate the necessary discussion of the ethics of document design, therefore, I devised a survey of seven questions regarding various document design situations. In this survey, each question is assessed on a five-point scale and a brief explanation of each answer is solicited (see Figure 10.1). Question 1 (the resume) looks at the practice of shrinking type and leading to fit more information on a page. Question 2 (the pie chart) investigates graphic design using persuasive coloring. Question 3 (the photograph) focuses on pictorial illustrations and the manipulation of visual information. Question 4 (the evaluation) explores how page design serves to direct the audience's attention. Question 5 (the proposal) examines the practice of inflating a document's size by increasing type size and leading. Question 6 (the line graph) focuses on the design of graphic illustrations that violate the typical reader's expectations. Question 7 (the warranty) addresses the issue of typography and readability.

Notice that I offer no definition of the word *ethical*: instead of testing the ability of the respondents to apply a given definition of ethics, the survey tries to determine how the respondents themselves define the word ethical within the seven document design situations.

My objective was to devise an instrument that was sufficiently provocative to stimulate discussion, both in school and on the job, as well as relatively quick and easy to administer so that it was practical for both academic and professional environments. I also sought to distribute this survey nationally, thereby allowing respondents to compare their answers to the answers of a representative sampling of the technical communication profession: such a comparison could itself lead to additional discussion of the ethical issues raised by the survey. In addition, a national survey could identify points of consensus from which might arise guiding principles on the ethics of document design.

In a pilot testing of this survey (Dragga 1993), I examined the opinions of practicing technical communicators as well as technical communication majors and minors. I distributed the survey to 33 professional technical communicators from five Dallas organizations and to 31 technical writing majors and minors enrolled in a senior-level course in technical and professional editing at Texas Tech University. This pilot testing indicated that the survey was easily administered and effective at stimulating discussion of ethical issues.

The pilot testing, however, also revealed that students were tentative in judging the seven situations, preferring "mostly ethical" or "mostly unethical" as their answers, whereas the majority of professional communicators chose either "completely ethical" or "completely unethical" as their answers. As a technical communication teacher, therefore, I considered it especially important that my national survey investigate possible differences of opinion dividing educators and practicing writers and editors. Because I am responsible for the education of technical communication majors and minors, I require a clear picture of the profession: Am I teaching ethical principles that professional writers and editors espouse or oppose? Without this knowledge, I do my students a genuine disservice. I encourage their timidity and slow their transition to the working world by failing to prepare them for the ethical challenges they are likely to encounter on the job.

Characteristics of the Survey

In January and February of 1994, I surveyed technical communicators and technical communication teachers to determine their perspectives on the ethics of document design. I identified the population for this survey by using a membership list supplied by STC of 500 technical writers/editors and 500 technical writing teachers in the U.S. (divided proportionally by zip code to achieve a geographical distribution representative of the STC membership). In addition to the seven questions on ethics (see Figure 10.1) I asked respondents to identify the following:

- Primary occupation (educator, technical communicator)
- Sex (male, female)
- Years of professional experience (≤ 2 years, 3-5 years, 6-10 years, 11+ years)
- Level of education (< Bachelor's degree, Bachelor's degree, Master's degree, Doctorate)

I requested information on sex, professional experience, and level of education because I believed such characteristics could be pertinent to the ethics of document design. Several studies, for example, reveal possible differences in the way that males and females perceive and use visual information (Geary, Gilger, and Elliott-Miller 1992; Goldstein, Haldane, and Mitchell 1990; Olson and Eliot 1986; Peterson 1983; Togo and Hood 1992), and a variety of theorists claim that males and females adopt differing ethical perspectives (Code, Mullett, and Overall 1988; Gilligan 1982; Kittay and Meyers 1987; Larrabee 1993; Noddings 1984). The demographic categories of this survey duplicate those of STC's 1992 membership survey, thus permitting a comparative analysis of the two populations.

I anticipated that the brevity of the survey would encourage a high rate of response and thus yield a representative sampling of opinion. The surveys were mailed with a postage-paid envelope and a brief cover letter soliciting the recipient's cooperation.

A Consensus on Consequences

Of the 1,000 surveys mailed, I received 455 replies, a response rate of 45.5%. While 66% of the technical communicators answered the survey, only 20% of the educators did. Table 10.1 displays the demographic information.

Only 430 people identified their occupation, 420 their sex, and 443 their professional experience and level of education. Relative to the STC membership (Society for Technical Communication 1992), this population has more educators (24% versus 10%), more men (45% versus 38%), more advanced degrees (55% versus 35%), and more job experience (typically 11+ years versus 7 years).

Table 10.2 displays the survey findings. Notice the clear consensus on Question 1 (the resume), Question 3 (the photograph), and Question 5 (the proposal). On the remaining questions, opinion is divided. A clear majority, however, consider Question 3 (the pie chart) and Question 4 (the evaluation) either "mostly ethical"

Table 10.1 Characteristics of Survey Respondents

Primary Occupation of Respondents

	Occupation	Count	Percent
1	Educator	102	24
2	Communicator	328	76

Sex of Respondents

	Sex	Count	Percent
1	Male	188	45
2	Female	232	55

Professional Experience of Respondents

	Experience	Count	Percent
1	< 2 years	33	7
2	3–5 years	82	19
3	6–10 years	105	24
4	11+ years	223	50

Level of Education of Respondents

	Degree	Count	Percent
1	< Bachelor's	29	6
2	Bachelor's	172	39
3	Master's	171	39
4	Doctorate	71	16

Table 10.2 Survey Findings

	Resume (% of 455 answers)	Pie Chart (% of 452 answers)	Photograph (% of 453 answers)	Evaluation (% of 449 answers)	Proposal (% of 444 answers)	Line Graph (% of 452 answers)	Warranty (% of 451 answers)
Completely Ethical	88.0	38.7	2.9	30.5	68.9	9.1	14.4
Most Ethical	9.2	25.7	2.0	24.1	13.2	8.6	18.9
Ethics Uncertain	4.2	18.4	9.5	26.7	12.6	15.5	22.6
Mostly Unethical	0.2	9.7	9.5	12.9	2.3	33.8	29.7
Completely Unethical	0.4	7.5	76.1	5.8	2.7	33.0	14.4

Table 10.3 Survey Responses, Men Vs. Women

	Mean				
Group	**Question 1**	**Question 2**	**Question 4**	**Question 6**	**Question 7**
Men	1.274	2.484	2.645	3.973	3.276
Women	1.13	2.009	2.175	3.524	2.961

Table 10.4 Survey Findings, Men Vs. Women, Questions 4 and 6

Question 4 Evaluation			
	Answer	**Men**	**Women**
1	Completely Ethical	25.8%	34.6%
2	Mostly Ethical	20.4	27.2
3	Uncertain	26.9	26.8
4	Mostly Unethical	17.2	8.8
5	Completely Unethical	9.7	2.6
Question 6 Line Graph			
	Answer	**Men**	**Women**
1	Completely Ethical	7.0%	10.0%
2	Mostly Ethical	6.5	10.8
3	Uncertain	11.8	18.6
4	Mostly Unethical	31.7	38.1
5	Completely Unethical	43.0	22.5

or "completely ethical," while a majority judge Question 6 (the line graph) either "mostly unethical" or "completely unethical." Question 7 (the warranty) elicits a genuinely divided opinion, with equal minorities labeling it "completely ethical" and "completely unethical"; however, a plurality judge it "mostly unethical."

Analysis of the findings according to occupation, education, and professional experience reveals no statistically significant differences. Educators and technical communicators judge the seven situations virtually identically, as do all levels of education and job experience. On Questions 1, 2, 4, 6, and 7, however, the answers of men and women exhibit statistically significant differences (unpaired two-tailed *t* test, $p \leq .01$), with women consistently more lenient or men consistently more strict in their judgments (see Table 10.3).

Only twice, however, is this difference sufficient to cross the 5-point scale (see Table 10.4). On Question 4, a plurality of men (26.9%) answered "ethics uncertain" while a plurality of women (34.6%) answered "completely ethical," and on Question 6 a plurality of men (43%) answered "completely unethical" while a plurality of women (38%) answered "mostly unethical."

Of the 455 respondents, 402 (88%) offered explanations of one or more of their answers and 304 (67%) offered explanations for all seven of their answers, sometimes in considerable detail, often with more than one type of explanation for a single answer—a vivid indication of the survey's ability to stimulate discussion of ethical issues. In discussing their answers, men and women were equally cooperative: 89%

of the men and 88% of the women explained one or more of their answers, and 65% of the men and 68% of the women explained all their answers.

I considered the explanations important because I wanted to know not only what people would do in a given situation, but why—thus to gain insight on their thinking as well as their actions. I analyzed the explanations and later reviewed my analysis to verify its accuracy and consistency. I examined and classified a total of 3,267 explanations, identifying nine categories (see Table 10.5).

I tried initially to classify the explanations according to their implied philosophy, such as Aristotle's golden mean (i.e., vice in the extremes, virtue in moderation), Kant's categorical imperative (i.e., unconditional and universal obligations of conscience), and Mill's principle of utility (i.e., the greatest good for the greatest number). While I could from time to time decipher the philosophical basis of a given explanation, however, it was impossible to do so with sufficient frequency or genuine confidence. My categories, as a consequence, are necessarily cautious, focusing on the explicit words of the explanations and avoiding interpretation of implicit philosophical perspectives. That is, I classified the explanations according to their locution as opposed to their illocution (Austin 1962).

Table 10.6 displays the distribution of the nine types of explanations for each of the seven survey questions. The most frequent type of explanation is consequences: it is also the prevailing explanation for five of the seven questions, with specifications the favorite explanation for the remaining two questions. The explanations are similarly divided across the nine categories for men and women in spite of the statistically significant differences in their survey answers (see Table 10.7).

With the exception of *insufficient information*, the least favorite answer is *Reader's Responsibility*: its highest frequency is 11% on Question 7—a clear rejection of a writer-based or caveat emptor philosophy of technical communication. Nevertheless, the explanation *Writer's Responsibility* is also atypical: its highest frequency is 12% on Question 1—a failure to affirm the STC principle "Hold myself responsible for how well my audience understands my message."

Also given little attention are *Common Practices* and *Intentions*: their highest frequency is 20% and 17%, respectively, on Question 6. On Question 4, *principles* achieves its highest frequency, 15%.

The relative frequency of specific types of explanations, however, disguises the rarity with which individuals display a consistent guiding philosophy. Of the 304 respondents who explained all their answers, only 3 offer a single type of explanation: either *consequences* or *intentions*. Of the remaining respondents, 243 (80%) offer four or more different types of explanations for their answers and the explanations are often contradictory. Consider, for example, one respondent's explanation for Question 6: "It is the reader's responsibility to carefully review the material." Here is the same respondent's explanation for Question 7: "It is your job to emphasize the positive aspects of your company. However, as a communicator it is important to assist the reader in comprehension."

Discussion

Technical communicators and technical communication teachers, regardless of their education and job experience, adopt similar ethical views on the questions of this survey. The perspectives of men and women, however, display statistically

Table 10.5 Explanations of Ethical Decisions

Explanation	Definition	Example from survey
Common Practices	Explains that the design obeys or violates a common practice.	• "I'm used to this style of representing policies and warranties. It's the normal way we receive such information. We have learned to accept this." • "Normal practice in evaluations is to emphasize the positive. Any mention of negatives is really damning." • "This is a common practice."
Specifications	Cites the presence or absence of explicit design specifications or regulations.	• "The specifications stated on page—not a particular number of words." • "The prospective employer gave no guidelines for résumé design. Your font, font size, leading, kerning, etc., options are infinite." • "The client asked for a physical restriction by number of pages without specifying type size, leading, etc."
Reader's Responsibility	Declares that readers are responsible for deciphering the communication.	• "The reader of the graph has some obligation to check how the data is presented." • "The reader must be responsible for carefully evaluating the information." • "People are responsible for reading warranties and taking care of themselves! 'Let the buyer beware' is the credo of the business world."
Writer's Responsibility	Focuses on the writer's ability or obligation to design the communication appropriately.	• "The writer is being compensated to put his/her organization in the best light (or color) possible. This is being accomplished in the pie chart." • "Making information inaccessible isn't why I'm in this profession." • "As a technical communicator my purpose is to communicate information as accurately as possible."
Writer's Intentions	Assesses the writer's motivations.	• "Underlying motive is dishonest—wanting audience to misinterpret information by not reading or readily comprehending it." • "The purpose is to deceive because you are hoping that the viewer will not understand what he/she is saying." • "The intent is to deceive the reader and to lead him/her into ignoring important information."
Consequences	Emphasizes the positive or negative consequences of the design.	• "It changes the meaning of the results in a way the reader is not likely to discover." • "This could be construed as an advantage to the reader, having larger type, thus easier to read." • "This misleads the reader and does not factually represent the situation."
Judgments	Offers a conditional or unconditional evaluator of the design.	• "If I can include more information and remain within the standard of making a page visually pleasing and readable, I think this is OK. " • "As long as all the info is there, phrased clearly, purposely making it hard to read is obnoxious and unadmirable, but not illegal." • "This is poor document design."

Table 10.5 (continued) Explanations of Ethical Decisions

Explanation	Definition	Example from survey
Principles	Identifies a guiding principle or ideal regarding the design.	• "For an evaluation, something as parallel as qualifications/deficiencies should be presented with parallel designs." • "Charts should show the data with neutrality (in general). Good points may be highlighted (if that is the purpose of the chart), but negative points should not be described in a fashion that misleads." • "The warranty should not be offered unless it's formatted so people can read it."
Insufficient Information	Offers no decision because of insufficient information.	• "It depends on who has asked you to evaluate the employee for promotion and why you want to de-emphasize deficiencies." • "Depends on the size of the font used for the percentages and other characteristics of the graphic. A number of other factors could affect the 'perceived' size of each pie slice." • "Whether or not it is ethical depends on the product and the intended users of the product."

Table 10.6 Frequency of Explanations

	Résumé	Pie Chart	Photograph	Evaluation	Proposal	Line Graph	Warranty
Consequences	106	160	243	118	118	207	117
Specifications	177	2	6	3	132	4	11
Judgments	73	83	57	77	111	66	75
Common Practices	39	29	2	16	20	108	80
Writer's Intentions	15	58	74	31	15	94	73
Principles	14	24	38	63	43	19	36
Writer's Responsibility	57	43	2	47	30	12	18
Reader's Responsibility	6	28	2	25	4	32	54
Insufficient Information	2	14	8	29	3	2	12

significant differences, pointing to psychological and social issues as possibly more important influences on the ethical choices of individuals. This finding of "strict" men and "lenient" women could also support Gilligan's claim that men ordinarily adopt a principle of justice to guide their ethical decisions, whereas women are more likely to exercise or integrate a principle of caring (1982, 1987). Nevertheless, on this survey, men and women offer virtually identical explanations of their answers, emphasizing the positive or negative consequences of a specific design decision.

Table 10.7 Frequency of Explanations, Men versus Women

	Men (%)	Women (%)
Common Practices	9	9
Specifications	9	10
Reader's Responsibility	4	5
Writer's Responsibility	6	7
Writer's Intentions	11	11
Consequences	34	32
Judgments	16	17
Principles	8	7
Insufficient Information	3	2

The findings also indicate that practitioners and educators in the field of technical communication seem to have achieved consensus on the practice of shrinking type and leading to fit more information on a page (completely ethical) and on the manipulation of pictorial illustrations (completely unethical). A clear majority consider inflating type and leading to fit less information on a page, choosing colors for persuasive purposes, and using spacing to direct or divert the reader's attention to be ethical design practices. A majority consider graphic distortion unethical. Using typography to decrease readability, however, elicits a divided opinion, even though the practice opposes the earlier mentioned STC directive: "Hold myself responsible for how well my audience understands my message."

While individuals offer a variety of explanations for their ethical choices and thus display no single guiding philosophy, the totality of the survey answers and explanations do espouse a basic principle of ethical document design: The greater the likelihood of deception and the greater the injury to the reader as a consequence of that deception, the more unethical is the design of the document. If deception and injury are unlikely, the design choices are ethical. In determining likelihood and weighing the degree of resulting injury, the writer might consider several issues, including typical communication practices, professional responsibilities, and explicit specifications and regulations, as well as rhetorical intentions and ideals.

This is essentially a "goal-based" philosophy: that is, "the rightness or wrongness of an action is a function of the goodness or badness of its consequences" (Wicclair and Farkas 1984, p. 15). It is also basically a negative guideline, emphasizing practices to avoid.

Is a philosophy that emphasizes consequences a sufficient or satisfactory basis of ethical decisions? If I decide the ethics of a situation according to its consequences, am I ethically obliged to weigh all the consequences? Consider Question 5 (the inflated proposal): none of the people citing consequences to justify their answers mention the ecological consequences of the scenario: that is, by inflating the proposal from 21 pages to 25 pages, the writer is obviously using paper unnecessarily. In a world of limited resources, is this ethical? Is it only the direct and immediate consequences that are important? Which consequences does the technical communicator ignore? Are the consequences to the writer or to the profession unimportant? Do writers jeopardize their credibility by exercising the little deceptions of Question 2 (the pie chart), Question 4 (the evaluation), or Question 7 (the warranty)? Does success with a little deception encourage a writer to practice bigger deceptions? Do such practices damage the reputation of all technical communicators? Is it always possible for individual writers on the job to perceive the direct and indirect consequences of their design decisions or to judge which consequences are important and which are unimportant? Is this expectation itself impractical and unethical?

Technical communicators thus seem to operate in isolation, without a guiding philosophy that genuinely guides, that espouses the considered opinion of the profession regarding ethical communication, a considered opinion achieved and disseminated through a comprehensive discussion of the technical communicator's several obligations—to himself or herself, to his or her organization, to the audience, to the subject, to the profession, and to society. Without this principle of "considered practice" to guide their decisions regarding document design, technical communicators have the virtually impossible job of continuously adapting their individual ethical practices to the rapid advances of computerized technology and the new rhetorical powers that such advances never cease to offer.

Conclusions and Recommendations

As I mentioned at the beginning of this article, the ethics of document design has received little investigation from technical communicators. Ideally, this survey and its tentative findings will encourage more exploration of this important topic. Specifically, additional research is necessary on other issues of document design such as the implications of line length, italics, white space, or the size and position on the page of illustrations. More study is needed to determine whether other document design issues or other ethical questions elicit different types of explanations or guiding principles. Also essential is research to verify or qualify the differences between men and women identified in this survey. Do men and women differ in their evaluation of other document design issues? Are women more lenient or men more strict in judging other ethical issues? What social or psychological factors might contribute to this difference?

The high response rate to this survey and the enthusiasm of the respondents in explaining their answers are indicators of the perceived importance of this subject and the efficacy of this survey in raising ethical questions and stimulating discussion. Similar instruments might be devised for addressing the ethics of invention, arrangement, and style. Especially important is the national distribution of the survey, allowing technical communicators to judge their answers to the questions relative to a cross-section of the profession. For example, I imagine that the respondents who cited "reader's responsibility" to justify their ethical choices would be surprised to discover the rarity of this explanation on the survey and might thus be motivated to review or revise their principles and practices. I recommend that STC periodically distribute brief surveys of ethical questions and report the results of such surveys to the membership to encourage a continuing examination of ethical issues. I also recommend that the surveys be distributed internationally to permit investigation of cultural differences regarding professional ethics.

A widely distributed survey, however, has its limitations. In this survey, for example, the written explanations of answers proscribed a reliable interpretation of the implied philosophical perspectives that guided ethical choices. The explanations were thus classified according to their explicit wording as opposed to their implicit meaning. A more intensive investigation would be possible if the questions of a survey could also be addressed during personal interviews with a restricted but representative population of teachers and practitioners.

In addition, STC could review its "Code for Communicators," soliciting commentary from the membership and encouraging a comprehensive analysis of ethical issues. For example, STC might consider revising its directive "Hold myself responsible for how well my audience understands my message" to give it either more specificity or more emphasis. This principle, though pertinent to the ethics of document design, fails to serve as a consistent ethical guideline for technical communicators and technical communication teachers.

I would also encourage the Association of Teachers of Technical Writing (ATTW) to address (through its journal, newsletter, publication series, and e-mail list) its ethical obligation to teach the ethics of technical communication. Is it ethical to teach the techniques and principles of document design without also discussing the ethics of document design? If level of education has little or no influence on the ethical decisions of technical communicators, is it because teachers ignore the subject of ethics? If level of professional experience also has no impact on ethical decisions, is it because technical communicators were never taught to consider the ethical implications of their rhetorical power?

People ordinarily like to think of themselves as ethical. I'm no exception. But I also imagine that it is all too often easy for each of us to get caught up in the immediate needs of the organizations for which we work, to feel the pressures of personal ambition, to do that which is convenient, to want whatever it takes to satisfy the boss or client while completing the job on time and within budget, and to rationalize the dubious practices we momentarily adopt. Periodic self-examination is thus important as a way of orienting ourselves again as professionals and reaffirming the principles of ethical communication. Quite possibly the most ethical thing we can do as a profession is to nourish the ongoing discussion of ethical issues.

Acknowledgments

I am grateful to the STC Research Grants Committee for its generous support of this project; to four anonymous reviewers for their perceptive and detailed advice; and to the members of STC who participated in this survey for sharing with me their time and their ideas about ethics.

References

Aristotle. 1980. *The Nicomachean Ethics*. Trans. David Ross. New York, NY: Oxford University Press.

Austin, J. L. 1962. *How to Do Things with Words*. Cambridge, MA: Harvard University Press.

Benson, Phillipa J. 1985. Writing visually: Design considerations in technical publications. *Technical Communication* 32:35-39.

Bryan, John. 1992. Down the slippery slope: Ethics and the technical writer as marketer. *Technical Communication Quarterly* 1:73-88.

Buccholz, William James. 1989. Deciphering professional codes of ethics. *IEEE Transactions on Professional Communication* 32:62-68.

Clark, Gregory. 1987. Ethics in technical communication: A rhetorical perspective. *IEEE Transactions on Professional Communication* 30:190-195.

Code, Lorraine, Sheila Mullett, and Christine Overall, eds. 1988. *Feminist Perspectives: Philosophical Essays on Methods and Morals.* Toronto, ON: University of Toronto Press.

Dragga, Sam. 1992. Evaluating pictorial illustrations. *Technical Communication Quarterly* 1:47-62.

———. 1993. The ethics of delivery. In *Rhetorical Memory and Delivery: Classical Concepts for Contemporary Composition and Communication*, ed. John Frederick Reynolds, 79-95. Hillsdale, NJ: Erlbaum.

Felker, Daniel B., Frances Pickering, Veda Charrow, V. Melissa Holland, and Janice C. Redish. 1981. *Guidelines for Document Designers.* Washington, DC: American Institutes for Research.

Geary, David C., Jeffrey W. Gilger, and Barbara Elliott-Miller. 1992. Gender differences in three-dimensional mental rotation: A replication. *Journal of Genetic Psychology* 15:115-118.

Gilligan, Carol. 1982. *In a Different Voice: Psychological Theory and Women's Development.* Cambridge, MA: Harvard University Press.

———. 1987. Moral orientation and moral development. In *Women and Moral Theory*, ed. Eva Feder Kittay and Diana T. Meyers, 19-33. Totowa, NJ: Rowman & Littlefield.

Girill, T. R. 1987. Technical communication and ethics. *Technical Communication* 34:178-179.

Goldstein, David, Diane Haldane, and Carolyn Mitchell. 1990. Sex differences in visual-spatial ability: The role of performance factors. *Memory and Cognition* 18:546-550.

Johannesen, Richard L. 1990. *Ethics in Human Communication*, 3rd ed. Prospect Heights, IL: Waveland Press.

Kant, Immanuel. 1964. *Groundwork of the Metaphysic of Morals.* Trans. H. J. Paton. New York, NY: Harper & Row.

Kittay, Eva Feder, and Diana T. Meyers, eds. 1987. *Women and Moral Theory.* Totowa, NJ: Rowman & Littlefield.

Kostelnick, Charles. 1990. The rhetoric of text design in professional communication. *The Technical Writing Teacher* 17:189-202.

Larrabee, Mary Jeanne, ed. 1993. *An Ethic of Care: Feminist and Interdisciplinary Perspectives.* New York, NY: Routledge.

Mill, John Stuart. 1957. *Utilitarianism.* Ed. Oskar Piest. Indianapolis, IN: Bobbs-Merrill.

Monmonier, Mark. 1991. *How to Lie with Maps.* Chicago, IL: University of Chicago Press.

Murch, Gerald M. 1985. Using color effectively: Designing to human specifications. *Technical Communication* 32:14-20.

Noddings, Nel. 1984. *Caring: A Feminine Approach to Ethics and Moral Education.* Berkeley, CA: University of CA Press.

Olson, Darlene M., and John Eliot. 1986. Relationships between experiences, processing style, and sex-related differences in performance on spatial tests. *Perceptual and Motor Skills* 62:447-460.

Perica, Louis. 1972. Honesty in technical communication. *Technical Communication* 15:2-6.

Peterson, Becky K. 1983. Tables and graphs improve reader performance and reader reaction. *Journal of Business Communication* 20 (2): 47-55.

Plunka, Gene A. 1988. The editor's nightmare: Formatting lists within the text. *Technical Communication* 35:37-80.

Radez, Frank. 1980. STC and the professional ethic. *Technical Communication* 27:5-6.

Rubens, Philip M. 1981. Reinventing the wheel? Ethics for technical communicators. *Journal of Technical Writing and Communication* 11:329-339.

Sachs, Harley. 1980. Ethics and the technical communicator. *Technical Communication* 27:7-10.

Schriver, Karen A. 1989. Document design from 1980 to 1989: Challenges that remain. *Technical Communication* 36:316-331.

Shimberg, H. Lee. 1980. Technical communicators and moral ethics. *Technical Communication* 27:10-12.

Society for Technical Communication. 1992. *Profile 92: STC Special Report*. Arlington, VA: Society for Technical Communication.

Togo, Dennis F., and Jacqueline H. Hood. 1992. Quantitative information presentation and gender: An interaction effect. *Journal of General Psychology* 119 (2): 161-168.

Tufte, Edward R. 1983. *The Visual Display of Quantitative Information*. Cheshire, CT: Graphics Press.

Walzer, Arthur E. 1989. Professional ethics, codes of conduct, and the Society for Technical Communication. In *Technical Communication and Ethics*, ed. R. John Brockmann and Fern Rook, 101-105. Arlington, VA: Society for Technical Communication.

White, Jan V. 1982. *Editing by Design*, 2nd ed. New York, NY: Bowker.

Wicclair, Mark R., and David K. Farkas. 1984. Ethical reasoning in technical communication: A practical framework. *Technical Communication* 31:15-19.

Williams, Robert I. 1983. Playing with format, style, and reader assumptions. *Technical Communication* 30:11-13.

Commentary

Sam Dragga's article is generally a good example of a report of survey results, though there are some problems with the survey's design and the presentation of the results. Despite—and to some extent, because of—the problems, the article provides some very interesting characteristics for discussion here.

The subject of ethics in technical communication is an important one, but few people would likely think about potential ethical problems involved in document design, instead focusing on more general business issues such as conflict of interest or honesty in disclosure of product flaws. As Dragga introduces his seven miniature case studies, however, we realize that document design can definitely be an ethical briar patch.

Purpose and Audience

Dragga's first sentence succinctly states his purpose: "This article reports the results of a national survey of [U.S.] technical communicators and technical communication teachers regarding their perspectives on the ethics of various document design scenarios." This sentence also suggests his primary research question: What are the opinions of practitioners and academics regarding the ethical appropriateness of certain document design choices?

As with the other articles we have examined, Dragga's intended audience is a mix of technical communication practitioners and academics, the two major components of the membership of the Society for Technical Communication who comprise its journal's chief audience. In fact, if we need to be reminded of this dual audience more explicitly, Dragga tells us at the beginning of the "Characteristics of the Survey" section that he sent his questionnaire to a large number of people from each of these two groups.

Organization

The report's organization is typical of many research reports:

- The "Introduction" section announces the article's purpose and asks readers to answer the survey questions themselves before continuing with the article.

- The article's second section, "Rhetorical Power and Ethical Obligations," is essentially a literature review and reports that almost nothing in the technical communication ethics literature addresses document design and very little in the document design literature that addresses ethics.
- "A Survey of Opinion on Ethics" describes the design of the survey questions designed to probe respondents' opinions.
- The fourth section, "Characteristics of the Survey," explains how the survey population was identified and details the demographic questions posed. This section also mentions previous research about possible gender differences in the perceptions of visuals and ethical perspectives.
- "A Consensus on Consequences" relates the findings from the survey's rating questions as well as some information about responses to the open-ended requests for explanations of rating choices.
- The "Discussion" section analyzes the findings, concentrating on respondents' open-ended explanations of their ratings.
- "Conclusions and Recommendations" encourages further exploration of the ethics of document design and suggests how organizations such as STC and the Association of Teachers of Technical Writing might respond to the findings reported here.
- At the end of the article are acknowledgments of the STC research grant that supported his study, the feedback supplied by the journal's anonymous peer reviewers, and the STC members who participated in the survey, as well as citations of all the works referenced in the text.

There are no surprises here in terms of organization.

Survey Construction

Information Elicited by the Questions. As we saw in Chapter 6, surveys can do several things: collect data, measure self-reported behavior, and assess attitudes. The questions in Dragga's survey attempt to do two of these.

- Seven questions elicit respondents' attitudes or opinions about the document design scenarios they presented.
- Four demographic questions seek information about the survey population that could be compared to data collected in STC's membership survey for 1992, and that would allow the researcher to determine whether gender, age, educational level, or occupation correlated in any significant way to the respondents' answers to the opinion questions.

Design and Structure of the Questionnaire

Question Design. Dragga has generally designed his opinion questions very well, though experts will quibble with the formulation of two of the implied demographic questions.

All of the questions successfully avoid absolute statements. They do not use terms such as *always* and *never*, and are all based on a particular scenario presented within the context of the question. Thus, the questions are well grounded in the specific, and avoid generalities.

Similarly, the opinion questions avoid the problem of statements in the negative. Each of them is identical: "Is this ethical?" And although we are not given the precise form of the demographic questions, we can assume from Table 1 that these questions did not contain statements in the negative. Therefore, we can state that the questions are stated in such a way that they are reasonably easy to understand and would not be confusing to the intended respondent group.

The questions are all unbiased. For example, although some of the opinion questions seem to present design choices that could have easily been portrayed as clearly unethical (for example, the line graph scenario in Question 6), the design decision is explained as an attempt to give a positive initial impression of the company rather than to deceive the reader.

Finally, each question focuses on a single concept: identifying the respondent's occupation or level of education, increasing or decreasing the amount of information on a page, emphasizing or de-emphasizing information, and so forth.

Question Types. The survey uses a mix of question types: Rating questions to identify respondents' views of the ethics of each of the seven scenarios, open-ended questions to probe respondents' reasons for assigning those ratings, and multiple-choice questions to gather demographic information.

Each of the opinion questions has two parts. The first part briefly presents a situation and asks the respondent to rate the ethics of the document design choice on a five-point semantic differential scale:

1. Completely ethical
2. Mostly ethical
3. Ethics uncertain
4. Mostly unethical
5. Completely unethical

As we mentioned in Chapter 6, rating questions have the advantage of helping to clarify respondents' attitudes and preferences, are easy for respondents to answer, and are easy for researchers to summarize. Because this survey was mailed to a large survey population, the ease of summarizing responses to these questions is quite important, but the other two characteristics are notable as well.

The questions ask survey respondents to put themselves in a series of situations where the ethical choices are not obvious or easy to make. For example, in Question 3, it is clear that there is a degree of dishonesty involved in asking an able-bodied employee to sit in a wheelchair for a picture to be used in a recruiting brochure, but is that ethical problem obviated by the message that the picture conveys—that those with disabilities are welcome to apply? By giving respondents three additional choices along an ethical scale between the two extremes, Dragga makes it easier for respondents to reply. Similarly, by providing degrees of agreement and disagreement with the ethical choices described in the scenario, the questions help respondents clarify their opinions about those choices.

The second part of each of the opinion questions is open-ended, asking respondents to explain why they chose the rating for each situation. These open-ended questions provide respondents with a chance to clarify the reasoning behind their ratings and help reduce the opportunity for researcher bias, but they significantly increase the difficulty of summarizing and reporting the results because each response will require some degree of interpretation by the researcher.

One of the disadvantages of open-ended questions, the demand on respondents' time, probably had both positive and negative effects on Dragga's research project. The request for explanations may have reduced the response rate among the survey population, particularly for the educators, whose response rate was poor. At the same time, the fact that "89% of the men and 88% of the women explained one or more of their answers, and 65% of men and 68% of women explained all their answers" suggests that the overwhelming majority of those interested enough to reply to the survey thought that it was important to explain the thinking behind their ethical choices and to take the additional time required to do so.

Dragga does not reproduce the four multiple-choice demographic questions from his survey in the article, but the responses to these questions are summarized in Table 10.1. As we pointed out in Chapter 6, multiple-choice questions provide an array of choices from which the respondent selects one or more answers. Each of Dragga's four demographic questions is designed to elicit a single response regarding the respondent's primary occupation, gender, years of professional experience, and level of education. However, two of the demographic questions raise problems.

The choices provided for professional experience are

- ≤ 2 years
- 3–5 years
- 6–10 years
- 11+ years

Dragga has selected these answers so that he can compare his demographic results with the results of STC's membership survey published in *Profile 92*, listed in his references at the end of the article. Unfortunately, either the STC survey or the report of its results was badly designed. How should someone with between 2 and 3 years, 5 and 6 years, or 10 and 11 years of experience respond to this question? Should they round up or down? The ambiguity in the available choices may have confused some respondents, perhaps explaining why two people did not respond to this question and perhaps making the responses of others less than accurate.

Similarly, the question about level of education attained may have been ambiguous to some respondents because it does not specify the major field of the various degree choices. For example, what answer should a respondent with a bachelor's degree in technical communication and a master's degree in business administration give? Again, such ambiguity may have resulted in respondent confusion, and made responses inaccurate and difficult to analyze. However, in both of these cases, the effect of the ambiguity was probably slight.

Exercise 10.1

Using the data that Dragga provides in Table 10.1, reconstruct the survey questions he used to collect the demographic information he reports. Correct the problems of ambiguity in two of the questions in such a way that demographic results from these questions can still be compared to those in the STC membership survey. ■

Survey Structure. Dragga tells us that "The surveys were mailed with ... a brief cover letter soliciting the recipient's cooperation." He does not reproduce the letter or any survey instructions, so we do not know whether they explained the purpose of the study, measures that would be taken to protect respondents' privacy, and the fact that participation was voluntary. The last point is probably moot because the survey population obviously chose to reply or not.

We also do not know the order in which questions were asked. We can assume that the opinion questions were posed in the order reproduced in Figure 1, but Dragga does not explicitly tell us whether the demographic questions were asked after the opinion questions or before. Again, it might be reasonable to suppose that the demographic questions were asked at the end of the survey, but we do not know for certain.

On the other hand, it seems apparent that the demographic questions were not mixed in with the opinion questions, and thus the survey did group the questions logically.

The fact that Dragga included a postage-paid return envelope made it easy for respondents to reply. From our current perspective more than 10 years later, it is important to remember that this survey, mailed in the first two months of 1994, was distributed at a time when many technical communicators and educators did not have access to e-mail or to the Web. "Snail mail" was the only means of reaching a large survey population such as that Dragga targeted. Today, the same survey could be conducted at virtually no cost and probably with a significantly higher return rate using an e-mail message in place of the letter and a link to a Web-based survey.

Survey Testing. Dragga's decision to pilot test his survey is commendable. Although we do not discuss pilot testing in Chapter 6, this practice is similar to piloting a usability test. It allows researchers to evaluate the test instrument and tweak it to ensure that it elicits the results that they are looking for.

In this case, Dragga tested with both technical communicators and advanced technical communication undergraduate majors and minors. This population reasonably approximates the practitioner population chosen for the full-scale survey, but it included no educators. Given the poor response rate from the educator population for the survey, we might wonder whether piloting with a sample of that population might have resulted in improvements that would have increased the response rate.

Dragga reports that as a result of the pilot, he realized that students were tentative in their opinions and "therefore, I considered it especially important that my national survey investigate possible differences of opinion dividing educators and practicing writers and editors." Does he mean that his original intention was to survey only practitioners and that he decided to include educators in the survey population as a result of the pilot? The answer to this important question cannot be answered from the article.

As he relates in the first paragraph of "Characteristics of the Survey," Dragga selected his survey population of 500 U.S. communicators and 500 U.S. educators from a membership list supplied by STC. As the society's population was approximately 90% practitioners and 10% educators, however, it is not clear why Dragga chose equal numbers of both groups for the survey population. Had he selected 900 communicators and 100 teachers, his sample population would have been more representative of the entire STC U.S. population.

Report of Results

Sam Dragga analyzed the responses to his multiple-choice demographic questions and his ranking opinion questions using some of the quantitative research techniques that we discussed in Chapter 4. He looked for correlations between demographic characteristics and the ethical rankings assigned by respondents, and found that the results showed no correlation between rankings and occupation, level of education, or experience, but a significant correlation between rankings and gender for most questions, with women tending to be more lenient than men. He also calculated the mean response for men and for women for each of the five questions where he found a statistically significant difference in response by gender.

Dragga also examined the responses to the open-ended requests for explanations of the ethical rankings using a qualitative research technique described in Chapter 5. In analyzing the explanations provided, he found that they fell into nine categories, and he therefore concluded that respondents were not directed by "a single guiding philosophy" in their judgments. Indeed, this is the most important conclusion of Dragga's article, leading to a recommendation that STC "encourag[e] a comprehensive analysis of ethical issues" and "revis[e] its directive 'Hold myself responsible for how well my audience understand my message' to give it either more specificity or more emphasis." He also recommends that the Association of Teachers of Technical Writing to teach the ethics of document design.

Finally, Dragga uses frequency distribution, discussed in Chapter 6.

- Table 10.1 reports the frequency of responses to each of the four demographic questions.
- Table 10.2 indicates the distribution of rankings for each of the seven questions.
- Table 10.4 shows the range of responses for both genders on the two questions where there was no significant difference by gender.
- Table 10.6 illustrates the frequency of each category of explanation for each of the opinion questions, and Table 10.7 shows the same frequency by gender.

Measures of Rigor. The quantitative results Dragga reports certainly meet the standards of rigor described in Chapter 4. However, he does not address all of the measures of rigor for surveys. Although he does reveal the response rate for the survey, there are some problems with that rate. Also, his article does not provide the margin of error for frequency distributions or confidence intervals for means.

Response Rate. As we have seen, Dragga tells us that he mailed his survey to 500 educators and 500 practitioners; 455 people replied, of whom 430 identified their occupations. We recall from Chapter 6 that response rate gives an indication about how representative the returned surveys are likely to be of the population surveyed.

Occupation	Number of Surveys Returned	Response Rate (% Surveys Returned)	Response Characterization
All	455	45.5	Not adequate
Educator	102	20.4	Not adequate
Practitioner	328	65.6	Good

As the table shows, the percentage of all surveys returned is not adequate to predict with confidence that the responses are representative of the 1,000 people to whom the survey was sent. Similarly, the percentage of surveys returned by educators is not adequate to predict with confidence that their responses are representative of the 500 educators to whom the survey was sent. However, the 65% response rate of the practitioners who returned the survey provides us with a good degree of confidence that their opinions are representative of the 500 practitioners to whom the survey was sent.

How might Dragga have increased the response rate for his survey? There seems to be no question that the length of the survey is reasonable and that the instrument is usable, despite our minor quibbles about two of the demographic questions. And it would certainly seem that the opinion questions themselves piqued the interest of the survey population.

Dragga does not mention offering an incentive (such as being entered in a drawing for a book, software package, gift certificate, or some other prize) to respondents, and doing so might have resulted in more surveys being returned. Another factor that might be relevant here (though we did not discuss it in Chapter 6) is timing. Dragga mailed his surveys in January and February, a time when educators are typically just beginning a new semester or finishing a quarter. These are times when teachers are typically much busier than usual and perhaps less likely to respond to a survey. Had he sent the survey in March, the response rate for educators might have been larger.

Margin of Error. As we noted in Chapter 6, when a researcher reports survey responses as frequency distributions and the researcher wants to infer that the results from the survey sample are representative of the population of interest, then the report should include the margin of error. The margin of error indicates the reliability of the inference based on the size of the sample and the level of confidence the researcher wants to have. The margin of error is a percentage value that lets the researcher say, "I think the real value in the population falls within the reported value plus or minus this margin of error." The confidence level is an indication of how reliable that statement is.

As we have seen, Dragga uses frequency distributions in his report, but he does not report the margin of error. If he had reported the margin of error at a 95% confidence level, here are the results based on the same three response rates we determined above.

Occupation	Sample Size	Margin of Error (95% confidence)
All	455	±4.6
Educator	102	±9.7
Practitioner	328	±5.4

Thus, the margins of error for all respondents and for practitioners are probably acceptable. After all, the political polls we see routinely have margins of error of 3 or 4 points, and these figures are generally predictive of electoral results except when results are very close. For educators only, however, the margin of error is certainly unacceptable because a nearly 20 point spread is not likely to be much help in predicting the opinion of the population of educators as a whole.

Indeed, as Dragga notes, the responses to his demographic questions show significant differences from the demographics of the overall STC population in 1992:

> ... this population has more educators (34% versus 10%), more men (45% versus 38%), more advanced degrees (55% versus 35%), and more job experience (typically 11+ years versus 7 years).

It is important to note that despite the use of frequency distributions in his article, Dragga never makes the claim that the results he reports are representative of STC or of the larger population of technical communicator practitioners and educators. However, careless readers might jump to that conclusion, so it would have been preferable for Dragga to specify the margin of error for his quantitative results. That would have given the reader fair warning that the overall margin of error is ±4.6%, a nearly 10-point spread.

Confidence Intervals. Dragga provides the mean survey responses of men versus women for the five questions where he found a statistically significant difference by gender in Table 10.3, but he does not provide us with a confidence interval that would allow us to infer that the mean results for the men and women sampled are representative of the general STC population of U.S. men and women. And because he does not provide the standard deviation for these means, we are not able to calculate the confidence intervals for ourselves.

Again, it is important to note that Dragga makes no claim that his results are representative of the larger population. Therefore, he is not obliged to report the confidence levels or provide the standard deviations for the means he supplies so that we can calculate them.

Although there are some methodological flaws, Dragga's article is a pioneering investigation of the ethical choices that often face document designers. This important article deserves the kind of follow-up studies that the author calls for in his "Conclusions and Recommendations."

Summary

This chapter has expanded on the concepts and methods presented in Chapter 6 by examining an article-length report of the results of a survey on ethical choices involving document design. Following the full text of the article, we have explored how its author went about writing it, examining its purpose, audience, and organization. We also looked carefully at the design of the survey questions and the selection of the survey population. Finally, we discussed the survey report in terms of measures of rigor: response rate, margin of error, and confidence interval. Exercises in the chapter provided practice in writing survey questions and in analyzing a survey's measures of rigor.

References

Dragga, S. 1996. "Is this ethical?" A survey of opinion on principles and practices of document design. *Technical Communication* 43:255–265.

Answer Key

Exercise 10.1

1. What is your primary occupation?
   - Communicator
   - Educator
2. What is your sex?
   - Male
   - Female
3. How many years of professional experience do you have in technical communication? (Round your answer up to the higher choice if you fall between two possible responses.)
   - 2 years or less
   - 3–5 years
   - 6–10 years
   - 11+ years
4. What is your level of education? (Indicate the highest level you have achieved regardless of major.)
   - Less than a bachelor's degree
   - Bachelor's degree
   - Master's degree
   - Doctorate

Index

C

D

E

F

G

H

I

J

K

L

M

N

O

P

Q

R

S

T

U

V

W

Made in the USA
Lexington, KY
12 September 2016